# FE Environmental Practice

Ashok V. Naimpally, PhD, PE
Kirsten Sinclair Rosselot, PE

PPI2PASS.COM
A **KAPLAN** COMPANY

> **Report Errors for This Book**
>
> PPI is grateful to every reader who notifies us of a possible error. Your feedback allows us to improve the quality and accuracy of our products. Report errata at **ppi2pass.com**.

**Digital Book Notice**

All digital content, regardless of delivery method, is protected by U.S. copyright laws. Access to digital content is limited to the original user/assignee and is non-transferable. PPI may, at its option, revoke access or pursue damages if a user violates copyright law or PPI's end-user license agreement.

**FE ENVIRONMENTAL PRACTICE**

Current release of this edition: 5

**Release History**

| date | edition number | revision number | update |
|---|---|---|---|
| Sep 2020 | 1 | 3 | Minor corrections. |
| Dec 2020 | 1 | 4 | Minor corrections. |
| Nov 2022 | 1 | 5 | Minor corrections. |

© 2022 Kaplan, Inc. All rights reserved.

All content is copyrighted by Kaplan, Inc. No part, either text or image, may be used for any purpose other than personal use. Reproduction, modification, storage in a retrieval system or retransmission, in any form or by any means, electronic, mechanical, or otherwise, for reasons other than personal use, without prior written permission from the publisher is strictly prohibited. For written permission, contact permissions@ppi2pass.com.

Printed in the United States of America.

PPI
ppi2pass.com

ISBN: 978-1-59126-636-5

# Table of Contents

**PREFACE AND ACKNOWLEDGMENTS** ................................................................... v

**ENGINEERING REGISTRATION IN THE UNITED STATES** ................................. vii

**COMMON QUESTIONS ABOUT THE FE ENVIRONMENTAL EXAM** ..................... xi

**HOW TO USE THIS BOOK** ..................................................................................... xiii

**NOMENCLATURE** ..................................................................................................... xv

**PRACTICE PROBLEMS**

    Problems ........................................................................................................ P1-1

    Solutions ...................................................................................................... P1-16

**PRACTICE EXAM 1**

    Problems ........................................................................................................ E1-1

    Solutions ...................................................................................................... E1-16

**PRACTICE EXAM 2**

    Problems ........................................................................................................ E2-1

    Solutions ...................................................................................................... E2-16

# Preface and Acknowledgments

*FE Environmental Practice* is intended to help engineers and students study effectively for the fundamentals of engineering (FE) environmental computer-based test (CBT), administered by the National Council of Examiners for Engineering and Surveying (NCEES).

One of the best ways to master engineering concepts is to practice solving problems. This book contains 109 practice problems and two full-length practice exams of 110 problems each. These resources—almost 330 original problems in total—mimic the format of the actual exam. Complete solutions are provided, so you can compare your work to the solving methods shown here.

The material presented here is intended to provide a review of topics most likely to be covered in the environmental engineering exam. Both the *NCEES Handbook* and the breakdown of problem topics provided by NCEES were used to achieve this end. Solving the problems contained in this book will refresh your knowledge and help you pass the exam.

Because examinees are only allowed to use the *NCEES Handbook* in the exam, *FE Environmental Practice* is designed to help familiarize students with the handbook as they study for the exam. Relevant sections and equations from the *NCEES Handbook* are called out when used (in the online version these are in blue). The environmental engineering exam contains problems that use many of the sections of the *NCEES Handbook*, not just the environmental engineering section. Familiarity with the *NCEES Handbook* will greatly aid students in passing the exam.

The authors would like to thank Dr. Sunil Hangal of RTP Environmental Associates, Inc. in Greenbrook, New Jersey for his review of draft materials, and Seda Ozel and Romit Ketkar, students at California State University, Long Beach, and Anil Acharya, for checking the solutions. The editorial staff at PPI, especially senior copy editor Tyler Hayes and publishing systems specialist Ellen Nordman, displayed enormous amounts of patience and provided many useful comments as the book was being prepared. Ashok would also like to thank his wife, Shobha, for her support and encouragement, and Kirsten would like to thank her family and friends for their cooperation and assistance.

Good luck with passing the exam and with all your future endeavors!

Ashok V. Naimpally, PhD, PE
Kirsten Sinclair Rosselot, PE

# Engineering Registration in the United States

## 1. ENGINEERING REGISTRATION

*Engineering registration* (also known as *engineering licensing*) in the United States is an examination process by which a state's board of engineering licensing (i.e., registration board) determines and certifies that you have achieved a minimum level of competence. This process protects the public by preventing unqualified individuals from offering engineering services.

Most engineers do not need to be registered. In particular, most engineers who work for companies that design and manufacture products are exempt from the licensing requirement. This is known as the *industrial exemption*. Nevertheless, there are many good reasons for registering. For example, you cannot offer consulting engineering design services in any state unless you are registered in that state. Even within a product-oriented corporation, however, you may find that registered engineers have more opportunities for employment and advancement.

Once you have met the registration requirements, you will be allowed to use the titles Professional Engineer (PE), Registered Engineer (RE), and Consulting Engineer (CE).

Although the registration process is similar in all 50 states, each state has its own registration law. Unless you offer consulting engineering services in more than one state, however, you will not need to register in other states.

### The U.S. Registration Procedure

To become a registered engineer in the United States, you will need to pass two written examinations. This exam covers basic subjects from the mathematics, physics, chemistry, and engineering classes you took during your first four university years. In rare cases, you may be allowed to skip this first exam.

The second exam is the *Principles and Practice of Engineering Exam*. The initials PE are also used. This exam is on topics within a specific discipline, and only covers subjects that fall within that area of specialty.

Most states have similar registration procedures. However, the details of registration qualifications, experience requirements, minimum education levels, fees, oral interviews, and examination schedules vary from state to state. For more information, contact your state's registration board (**ncees.org**).

### National Council of Examiners for Engineering and Surveying

The *National Council of Examiners for Engineering and Surveying* (NCEES) produces, distributes, and scores the national FE and PE exams. The individual states purchase the exams from NCEES and administer them themselves. NCEES does not distribute applications to take the exams, administer the exams or appeals, or notify you of the results. These tasks are all performed by the states.

### Reciprocity Among States

With minor exceptions, having a license from one state will not permit you to practice engineering in another state. You must have a professional engineering license from each state in which you work. For most engineers, this is not a problem, but for some, it is. Luckily, it is not too difficult to get a license from every state you work in once you have a license from one state.

All states use the NCEES exams. If you take and pass the FE or PE exam in one state, your certificate will be honored by all of the other states. Although there may be other special requirements imposed by a state, it will not be necessary to retake the FE and PE exams. The issuance of an engineering license based on another state's license is known as *reciprocity* or *comity*.

The simultaneous administration of identical exams in all states has led to the term *uniform examination*. However, each state is still free to choose its own minimum passing score and to add special questions and requirements to the examination process. Therefore, the use of a uniform exam has not, by itself, ensured reciprocity among states.

## 2. THE FE EXAM

### Applying for the Exam

Each state charges different fees, specifies different requirements, and uses different forms to apply for the exam. Therefore, it will be necessary to request an application from the state in which you want to become registered. You'll find contact information (websites, telephone numbers, email addresses, etc.) for all U.S. state and territorial boards of registration at **ncees.org**.

Keep a copy of your exam application, and send the original application by certified mail, requesting a delivery receipt. Keep your proof of mailing and delivery with your copy of the application.

## Exam Dates

The national FE exams and computer-based PE exams are administered year-round. The pencil-and-paper PE exams are offered twice a year (usually in mid-April and late October), on the same weekends in all states. Check www.ppi2pass.com for a current exam schedule.

## FE Exam Format

The NCEES Fundamentals of Engineering examination has the following format and characteristics.

- The exam is six hours long, with an optional 25 minute scheduled break.

- Examination questions are presented via computer. A digital version of the *NCEES Handbook* is made available to the examinee on the same computer.

- Formulas and tables of data needed to solve questions in the exams are found in either the *NCEES Handbook* or in the body of the question statement itself.

- The exam has 110 questions. Most questions are multiple-choice questions, each with four possible answers lettered (A) through (D). Some questions will ask the examinee to select multiple correct answers, or fill in a blank, or drag and drop images on the screen.

- Guessing is valid; no points are subtracted for incorrect answers.

- The number of questions for each subject in the environmental exam are given in Table 1.

## Use of SI Units on the FE Exam

Metric questions are used in all subjects, except some civil engineering and surveying subjects that typically use only customary U.S. (i.e., English) units. SI units are consistent with ANSI/IEEE standard 268 (the American Standard for Metric Practice). Non-SI metric units might still be used when common or where needed for consistency with tabulated data (e.g., use of bars in pressure measurement).

## Grading and Scoring the FE Exam

The FE exam is not graded on the curve, and there is no guarantee that a certain percentage of examinees will pass. Rather, NCEES uses a modification of the *Angoff procedure* to determine the suggested passing score (the cutoff point or *cut score*).

*Table 1* FE Environmental Exam Subjects

| subject | number of questions |
|---|---|
| Mathematics | 4–6 |
| Probability and Statistics | 3–5 |
| Ethics and Professional Practice | 5–8 |
| Engineering Economics | 4–6 |
| Materials Science | 3–5 |
| Environmental Science and Chemistry | 11–17 |
| Risk Assessment | 5–8 |
| Fluid Mechanics | 9–14 |
| Thermodynamics | 3–5 |
| Water Resources | 10–15 |
| Water and Wastewater | 14–21 |
| Air Quality | 10–15 |
| Solid and Hazardous Waste | 10–15 |
| Groundwater and Soils | 9–14 |

With this method, a group of engineering professors and other experts estimate the fraction of minimally qualified engineers who will be able to answer each question correctly. The summation of the estimated fractions for all test questions becomes the passing score. Because the law in most states requires engineers to achieve a score of 70% to become licensed, you may be reported as having achieved a score of 70% if your raw score is greater than the passing score established by NCEES, regardless of the raw percentage.

About 20% of the FE exam questions are repeated from previous exams—this is the *equating subtest*. Since the scores of previous examinees on the equating subtest are known, comparisons can be made between the two exams and examinee populations. These comparisons are used to adjust the passing score.

The individual states are free to adopt their own passing score, but all adopt NCEES's suggested passing score because the states believe this cutoff score can be defended if challenged.

You will receive the results within 12 weeks of taking the exam. If you pass, you will receive a letter stating that you have passed. If you fail, you will be notified that you failed and be provided with a diagnostic report.

## Permitted Reference Material

Since October 1993, the FE exam has been what NCEES calls a "limited-reference" exam. This means that no books or references other than those supplied by NCEES may be used. Therefore, the FE exam is really an "NCEES-publication only" exam. NCEES provides its own *FE Reference Handbook* (*NCEES Handbook*) for use during the examination. No books from other publishers may be used.

## 3. CALCULATORS

To prevent unauthorized transcription and redistribution of the exam questions, calculators with communication or text-editing capabilities are banned from all NCEES exam sites. You cannot share calculators with other examinees. For a list of allowed calculators check **ncees.org**.

It is essential that a calculator used for engineering examinations have the following functions.

- trigonometric functions
- inverse trigonometric functions
- hyperbolic functions
- pi
- square root and $x^2$
- common and natural logarithms
- $y^x$ and $e^x$

For maximum speed, your calculator should also have or be programmed for the following functions.

- extracting roots of quadratic and higher-order equations
- converting between polar (phasor) and rectangular vectors
- finding standard deviations and variances
- calculating determinants of $3 \times 3$ matrices
- linear regression
- economic analysis and other financial functions

## 4. STRATEGIES FOR PASSING THE FE EXAM

The most successful strategy for passing the FE exam is to prepare in all of the exam subjects. Do not limit the number of subjects you study in hopes of finding enough questions in your strongest areas of knowledge to pass.

Fast recall and stamina are essential to doing well. You must be able to quickly recall solution procedures, formulas, and important data. You will not have time during the exam to derive solutions methods—you must know them instinctively.

This ability must be maintained for six hours. Be sure to gain familiarity with the *NCEES Handbook* by using it as your only reference for some of the problems you work during study sessions.

In order to get exposure to all exam subjects, it is imperative that you develop and adhere to a review schedule. If you are not taking a classroom prep course (where the order of your preparation is determined by the lectures), prepare your own review schedule.

There are also physical demands on your body during the exam. It is very difficult to remain alert and attentive for six hours or more. Unfortunately, the more time you study, the less time you have to maintain your physical condition. Thus, most examinees arrive at the exam site in peak mental condition but in deteriorated physical condition. While preparing for the FE exam is not the only good reason for embarking on a physical conditioning program, it can serve as a good incentive to get in shape.

It will be helpful to make a few simple decisions prior to starting your review. You should be aware of the different options available to you. For example, you should decide early on to

- use SI units in your preparation
- perform electrical calculations with effective (rms) or maximum values
- take calculations out to a maximum of four significant digits
- prepare in all exam subjects, not just your specialty areas

At the beginning of your review program, you should locate a spare calculator. It is not necessary to buy a spare if you can arrange to borrow one from a friend or the office. However, if possible, your primary and spare calculators should be identical. If your spare calculator is not identical to the primary calculator, spend some time familiarizing yourself with its functions.

### A Few Days Before the Exam

There are a few things you should do a week or so before the exam date. For example, visit the exam site in order to find the building, parking areas, examination room, and rest rooms. You should also make arrangements now for child care and transportation. Since the exam does not always start or end at the designated times, make sure that your child care and transportation arrangements can tolerate a late completion.

Next in importance to your scholastic preparation is the preparation of your two examination kits. The first kit consists of a bag or box containing items to bring with you into the examination room.

- [ ] letter admitting you to the exam
- [ ] photographic identification
- [ ] main calculator
- [ ] spare calculator
- [ ] extra calculator batteries
- [ ] unobtrusive snacks
- [ ] travel pack of tissues
- [ ] headache remedy
- [ ] $2.00 in change
- [ ] light, comfortable sweater
- [ ] loose shoes or slippers
- [ ] handkerchief
- [ ] cushion for your chair
- [ ] small hand towel
- [ ] earplugs
- [ ] wristwatch with alarm
- [ ] wire coat hanger
- [ ] extra set of car keys

The second kit consists of the following items and should be left in a separate bag or box in your car in case you need them.

- [ ] copy of your application
- [ ] proof of delivery
- [ ] this book
- [ ] other references
- [ ] regular dictionary
- [ ] scientific dictionary
- [ ] course notes in three-ring binders
- [ ] instruction booklets for all your calculators
- [ ] light lunch
- [ ] beverages in thermos and cans
- [ ] sunglasses
- [ ] extra pair of prescription glasses
- [ ] raincoat, boots, gloves, hat, and umbrella
- [ ] street map of the exam site
- [ ] note to the parking patrol for your windshield explaining where you are, what you are doing, and why your time may have expired
- [ ] battery-powered desk lamp

## The Day Before the Exam

Take the day before the exam off from work to relax. Do not cram the last night. A good prior night's sleep is the best way to start the exam. If you live far from the exam site, consider getting a hotel room in which to spend the night.

Make sure your exam kits are packed and ready to go.

## The Day of the Exam

You should arrive at least 30 minutes before the exam starts. This will allow time for finding a convenient parking place, bringing your materials to the exam room, and making room and seating changes. Be prepared, though, to find that the examination room is not open or ready at the designated time.

Once the examination has started, consider the following suggestions.

Set your wristwatch alarm for five minutes before the end of each session, and use that remaining time to guess at all of the remaining unsolved problems. Do not work up until the very end. You will be successful with about 25% of your guesses, and these points will more than make up for the few points you might earn by working during the last five minutes.

Do not spend more than two minutes per question. (The average time available per problem is two minutes.) If you have not finished a question in that time, make a note of it and move on.

Do not ask your proctors technical questions. Even if they are knowledgeable in engineering, they will not be permitted to answer your questions.

Make a quick mental note about any problems for which you cannot find a correct response or for which you believe there are two correct answers. Errors in the exam are rare, but they do occur. Being able to point out an error later might give you the margin you need to pass. Since such problems are almost always discovered during the scoring process and discounted from the exam, it is not necessary to tell your proctor, but be sure to mark the one best answer before moving on.

# Common Questions About the FE Environmental Exam

Q: Will my FE certificate be recognized by other states?

A: Yes. All states recognize passing the FE exam.

Q: What is the format of the FE Environmental exam?

A: The FE Environmental exam is 6 hours long. There are 110 problems. The average time per problem is 2.5 minutes. Each problem is multiple choice with 4 answer choices. Some problems may be alternative item types. Most problems require the application of more than one concept (i.e., formula).

Q: Is there anything wrong with guessing?

A: There is no penalty for guessing. No credit is given for scratch pad work, methods, etc.

Q: Are the answer choices close or tricky?

A: Answer choices are not particularly close together in value, so the number of significant digits is not going to be an issue. Wrong answers, referred to as "distractors" by NCEES, are credible. However, the exam is not "tricky"; it does not try to mislead you.

Q: Are any problems related to each other?

A: Several questions may refer to the same situation or figure. However, NCEES has tried to make all of the questions independent. If you make a mistake on one question, it shouldn't carry over to another.

Q: Does the FE exam cover material at the undergraduate or graduate level?

A: Test topics come entirely from the typical undergraduate degree program. However, the emphasis is primarily on material from the third and fourth year of your program. This may put examinees who take the exam in their junior year at a disadvantage.

Q: Do you need practical work experience to take the FE Environmental exam?

A: No.

Q: Is the FE Environmental exam in customary U.S. or SI units?

A: The FE Environmental exam uses both customary U.S. and SI units.

Q: Does the *NCEES Handbook* cover everything that is on the FE Environmental exam?

A: No. You may be tested on subjects that are not in the *NCEES Handbook*. However, NCEES has apparently adopted an unofficial policy of providing any necessary information, data, and formulas in the stem of the question. You will not be required to memorize any formulas.

Q: Is everything in the *NCEES Handbook* going to be on the exam?

A: Apparently, there is a fair amount of reference material that isn't needed for every exam. There is no way, however, to know in advance what material is needed.

Q: How long does it take to prepare for the FE Environmental exam?

A: Preparing for the FE Environmental exam is similar to preparing for a mini PE exam. Engineers typically take two to four months to complete a thorough review for the PE exam. However, examinees who are still in their degree program at a university probably aren't going to spend more than two weeks thinking about, worrying about, or preparing for the FE Environmental exam. They rely on their recent familiarity with the subject matter.

Q: Where can I get even more information about the FE Environmental exam?

Visit the FE Exam Resource Center at **ppi2pass.com**

# How to Use This Book

*FE Environmental Practice* begins with a nomenclature list. The heart of this book consists of a chapter of nearly 110 practice problems and two 110-problem practice exams. These three sections are designed to guide your exam preparation by helping you determine the topic areas in which you need the most intensive review. These problems will refresh your memory of how to use the formulas and variables that are important in solving different problem types.

After you have worked the practice problems and reviewed any topics that need attention, take the first practice exam, using only the *NCEES FE Reference Handbook* (*NCEES Handbook*) as a reference. Detailed solutions appear at the end of the exam. Use them to evaluate your strengths and weaknesses. You might decide to study some topic areas in greater depth before taking the second practice exam. Again, when taking the second practice exam, use only the *NCEES Handbook* as a reference.

It is particularly important to get familiar with the environmental engineering section of the *NCEES Handbook*. However, because of the nature of the subject matter in the environmental engineering exam, much of the data and many of the formulas needed for solving problems appear in other sections of the handbook. While studying for the exam, be sure to use the *NCEES Handbook* as much as possible so that you'll know where to find specific formulas when you are taking the actual exam. *NCEES Handbook* section and equation names have been added throughout the solutions to point you to relevant content.

# Nomenclature

| | | |
|---|---|---|
| $a$ | empirical constant | $yd^3/in^2$ |
| $a$ | mass of adsorbed solute required to saturate a unit mass of adsorbent | g |
| $A$ | surface area | $m^2$ |
| $A$ | total collection area | $m^2$ |
| $A_{out}$ | concentration in effluent air | $kmol/m^3$ |
| $A_p$ | area of blade perpendicular to the direction of travel | $m^2$ |
| $A_{plan}$ | area of cross section of a packed bed | $m^2$ |
| AF | absorption factor | decimal |
| AT | averaging time | d |
| $b$ | empirical constant | $yd^3/lbf$ |
| $B$ | channel width for open-channel flow in a rectangular channel | m |
| BCF | bioconcentration factor | – |
| BW | body weight (body mass) | kg |
| $C$ | runoff coefficient | – |
| $C$ | roughness coefficient for Hazen-Williams equation | – |
| $C$ | concentration | decimal, mg/L, ppm, mol/$cm^3$, $\mu g/m^3$ |
| $C_{A,atm}$ | concentration of compound A at the surface of a landfill cover | $g/cm^3$ |
| $C_{A,fill}$ | concentration of compound A at the bottom of a landfill cover | $g/cm^3$ |
| $C_D$ | drag coefficient | – |
| $C_{max}$ | maximum ground-level concentration at ground level downwind from elevated source | $\mu g/m^3$ |
| $C_{org}$ | equilibrium concentration in organism | mg/kg or ppm |
| $C_{soil}$ | concentration of chemical in organic carbon component of soil | $\mu$g adsorbed/kg organic carbon or ppb |
| CE | combustion efficiency | % |
| CN | curve number | – |
| CO | volume concentration (dry) of CO | ppmv |
| $CO_2$ | volume concentration (dry) of $CO_2$ | ppmv |
| CSF | cancer slope factor | $(mg/kg \cdot d)^{-1}$ |
| CV | coefficient of variation | – |
| $d$ | diameter | m |
| $d$ | thickness of clay liner | m |
| $d_{pc}$ | diameter of particle that is collected with 50% efficiency | m |
| $D$ | depth | m |
| $D$ | diffusion coefficient | $cm^2/s$ |
| $D$ | dissolved oxygen deficit | mg/L |
| $D_d$ | cyclone's dust exit diameter | m |
| $D_e$ | cyclone's gas exit diameter | m |
| DO | dissolved oxygen concentration | mg/L |
| DRE | destruction and removal efficiency | % |
| $E$ | emission rate | kg/h |
| $E$ | specific energy in open-channel flow | m |
| $E$ | voltage requirement of electrodialysis unit | V |
| $E_1$ | removal efficiency during electrodialysis | – |
| $E_2$ | current efficiency during electrodialysis | – |
| ED | exposure duration | d |
| EF | exposure frequency | – |
| ET | amount of water lost through evapotranspiration per unit area | in |
| $f'$ | friction factor for packed bed | – |
| $f'_{ij}$ | friction factors for each media fraction | – |
| $f_{oc}$ | fraction of organic carbon in the soil | – |
| $F_A$ | average emission factor | kg/h/piece |
| $F$ | Faraday's constant | C/mol |
| $F$ | fraction of influent BODs consisting of raw primary sewage | – |
| $g$ | acceleration due to gravity | $m/s^2$ |
| $G$ | mixing intensity; root mean square velocity | m/s |
| $h$ | hydraulic head | m |
| $h$ | physical stack height | m |
| $\Delta h$ | plume rise | m |
| $h_f$ | head loss through packed bed | m |

| Symbol | Description | Units |
|---|---|---|
| $H$ | cyclone inlet height | m |
| $H$ | effective stack height | m |
| $H$ | hydraulic head | m |
| $H$ | Henry's law constant | Pa |
| $H'$ | dimensionless Henry's law constant | – |
| $H_L$ | head loss in mixing zone | m |
| HI | hazard index | – |
| $I$ | current during electrodialysis | A |
| $I$ | intake rate | mg/d |
| $I$ | rainfall intensity | in/hr |
| $J$ | flux through a membrane | mol/cm$^2 \cdot$s |
| $J_w$ | volumetric flux in ultrafiltration | m/s |
| $k$ | growth-rate fitting parameter | d$^{-1}$ |
| $k$ | rate constant | d$^{-1}$ |
| $k$ | treatability constant | 1/min |
| $k_1$ | constant in Hazen-Williams equation | – |
| $K$ | coefficient of permeability | cm/s |
| $K$ | constant for Manning's equation | – |
| $K$ | distribution coefficient | – |
| $K$ | experimental constant | – |
| $K$ | hydraulic conductivity | m/s |
| $K$ | partition coefficient | – |
| $K_d$ | distribution coefficient | – |
| $K_d$ | reaction-rate constant | d$^{-1}$ |
| $K_{La}$ | overall transfer-rate constant | 1/s |
| $K_p$ | mass transfer coefficient of solute across membrane | cm/s |
| $K_s$ | substrate concentration at which the specific growth rate is one-half of the maximum specific growth rate | g/m$^3$ |
| $K_T$ | impeller constant for turbulent flow | – |
| $L$ | length, depth | m |
| $L$ | liquid molar loading rate | kmol/s$\cdot$m$^2$ |
| $L$ | ultimate BOD (BOD remaining at time $t=\infty$) | mg/L |
| LADD | lifetime average daily dose | mg/kg$\cdot$d |
| $m$ | mass | kg |
| $M$ | molar density | kmol/m$^3$ |
| $M$ | sludge production rate | kg/d |
| $n$ | experimental constant | – |
| $n$ | Manning's roughness coefficient | – |
| $n$ | media characteristic coefficient | – |
| $n$ | number of cells | – |
| $n$ | number of moles | – |
| $n$ | rotational speed | rev/s |
| $N$ | gas flux | g/cm$^2 \cdot$s |
| $N$ | normality of solution | mol/L |
| $N$ | number of atoms | – |
| $N$ | radionuclide activity | Bq |
| $N_e$ | number of effective turns gas makes in cyclone | – |
| $p$ | overburden pressure | Pa |
| $P$ | population | persons |
| $P$ | power | ft$\cdot$lbf/sec |
| $P$ | precipitation per unit area | in |
| $P$ | sound pressure | Pa |
| $P$ | wetted perimeter | m |
| $\Delta P$ | net transmembrane pressure | Pa |
| $P_0$ | reference sound pressure | Pa |
| $P_v$ | volatile fraction of suspended solids | – |
| PER$_{sw}$ | amount of water per unit area percolating through landfill cover into compacted solid waste | in |
| $q$ | hydraulic loading | m$^3$/m$^2 \cdot$min |
| $q$ | unit discharge | m$^2$/s |
| $Q$ | emission rate | $\mu$g/s |
| $Q$ | runoff from rainfall per unit area | in |
| $Q$ | volumetric flow rate | m$^3$/s |
| $r$ | membrane pore size | m |
| $r_x$ | distance from source to receptor at point $x$ | m |
| $R$ | hydraulic radius | m |
| $R$ | recycle ratio | – |
| $R$ | electrical resistance | $\Omega$ |
| $R$ | retardation factor | – |
| $R$ | runoff per unit area | in |
| $R$ | stripping factor | – |
| $R$ | universal gas constant | J/mol$\cdot$K |
| Re | Reynolds number | – |
| RfD | reference dose | mg/kg$\cdot$d |
| $s$ | standard deviation of sample | various |
| $S$ | BOD | kg/m$^3$ |
| $S$ | concentration | g/mL |
| $S$ | cyclone vortex finder length | m |
| $S$ | maximum basin retention | m |
| $S$ | slope | m/m |
| $S$ | specific weight | kg/m |
| $S_0$ | initial BOD ultimate in mixing zone | mg/L |

| Symbol | Description | Units |
|---|---|---|
| $\Delta S_{LC}$ | change in amount of water held in storage in a unit volume of landfill cover | m |
| SPL | sound pressure level | dB |
| SW | specific weight | kg/m |
| $t$ | time | s |
| $t_c$ | critical time that corresponds to minimum dissolved oxygen | d |
| $t_{1/2}$ | half-life | d |
| $T$ | absolute temperature | K |
| $u$ | average wind speed at stack height | m/s |
| UF | uncertainty factor | – |
| v | velocity | m/s |
| $v_o$ | overflow rate | m/s |
| $V$ | volume | m³ |
| $V_1$ | raw sludge input | m³/d |
| $V_2$ | digested sludge accumulation | m³/d |
| $W$ | mass flow rate of a particular POHC | kg/h or lbm/h |
| $W$ | terminal drift velocity | m/s |
| $W$ | width | m |
| WF | weight factor | – |
| WOR | weir overflow rate | m²/s |
| $x$ | downwind distance along plume centerline | m |
| $x$ | mass of solute adsorbed | g |
| $\bar{x}$ | sample average | various |
| $X$ | concentration | ppb or µg/kg |
| $X$ | mass of adsorbed solute per unit mass of adsorbent | – |
| $y$ | depth of flow in open-channel flow | m |
| $y$ | horizontal distance from plume centerline | m |
| $y_t$ | amount of BOD exerted at time $t$ | mg/L |
| $Y$ | yield coefficient | – |
| $z$ | vertical distance from ground level | m |
| $Z$ | total carbon depth | m |
| $\Delta Z$ | membrane thickness | cm |
| $Z_s$ | depth of sorption zone | m |

## Symbols

| Symbol | Description | Units |
|---|---|---|
| $\alpha$ | kinetic energy correction factor | – |
| $\alpha$ | probability of getting a false positive | – |
| $\beta$ | probability of getting a false negative | – |
| $\gamma$ | specific weight of water | kg/m |
| $\delta$ | membrane thickness | m |
| | membrane porosity | % |
| $\eta$ | fractional particle collection efficiency | % |
| $\eta$ | porosity or effective porosity | % |
| $\eta$ | volumetric moisture content fraction of sediment or soil | kg/m³ |
| $\theta$ | hydraulic residence time | d |
| $\theta_c$ | solids residence time | d |
| $\mu$ | background concentration | various |
| $\mu$ | specific growth rate | 1/s |
| $\mu$ | viscosity of gas | kg/m·s |
| $\nu$ | number of ions formed from one molecule of electrolyte | – |
| $\nu_p$ | relative velocity of paddle | m/s |
| $\pi$ | osmotic pressure | Pa |
| $\rho$ | density | kg/m³ |
| $\rho_f$ | density of fluid | kg/m³ |
| $\sigma_y$ | horizontal dispersion parameter | m |
| $\sigma_z$ | vertical dispersion parameter | m |
| $\tau$ | half-life | s |
| $\Phi$ | osmotic coefficient | – |

## Subscripts

| | |
|---|---|
| $a$ | reaeration |
| $A$ | aeration basin or tank |
| $A$ | air |
| ave | average |
| $b$ | biological |
| $b$ | body |
| $B$ | background |
| $B$ | backwash |
| $B$ | breakthrough |
| $c$ | cone |
| $c$ | critical |
| $d$ | deoxygenation |
| $d$ | digester |
| $d$ | microbial death |
| $d$ | soil-water |
| $e$ | effective |
| $e$ | effluent |
| $e$ | equilibrium |
| fb | fluidized filter media |
| $g$ | gas |
| gas | gas-filled |
| $h$ | horizontal |
| $h$ | hydraulic |
| $i$ | impeller |
| $i$ | influent |
| $M$ | media |
| $o$ | effluent |
| $o$ | octanol phase |
| oc | organic carbon |
| org | organism |
| ow | octanol-water |
| $p$ | particle |
| $r$ | radioactive |
| $r$ | reaction |
| $R$ | recycle |
| $s$ | solute |
| $s$ | storage |
| $S$ | sample |
| sat | saturated |
| sw | soil-water |
| $t$ | terminal |
| $t$ | thickening |
| $T$ | time |
| $T$ | treated at exhaustion |
| TOC | total organic compounds |
| $w$ | aqueous phase, water |
| $w$ | waste sludge |
| $W$ | water |
| wi | initial compacted waste |
| wp | waste material at pressure $p$ |
| $x$ | cross-sectional |
| $x$ | particle |

# Practice Problems

**1.** A city has a population of 30,000. The ratio of the minimum daily flow of sewage to the peak flow rate of sewage is most nearly

(A) 0.16

(B) 0.24

(C) 0.40

(D) 0.60

**2.** The population of a town was 1.1 million in 1980, 1.2 million in 1990, and 1.3 million in 2000. The per-capita consumption of water was 0.5 kg/person/d in 1970 and 0.7 kg/person/d in 1990. The total water consumption in 2010 is expected to be

(A) $9.0 \times 10^5$ kg/d

(B) $1.2 \times 10^6$ kg/d

(C) $1.3 \times 10^6$ kg/d

(D) $1.4 \times 10^6$ kg/d

**3.** Dumping treated sewage discharge into oceans is

(A) not allowed presently for planning purposes for coastal towns

(B) allowed, but needs to be done at a distance of 12 mi from the shore

(C) allowed, but is best done using multiport diffusers from the bottom of the ocean

(D) allowed only by special permission from the EPA

**4.** A parcel of open land at the back of a residential area has an area of 50 000 m². If the open land is of the nature of parkland with a runoff coefficient of 0.25, the discharge due to a rainfall of 2.5 cm/h is most nearly

(A) 0.017 m³/s

(B) 0.087 m³/s

(C) 0.17 m³/s

(D) 0.87 m³/s

**5.** In a piping system, a 6 cm inside diameter pipe splits into two 10 m long pipes with inside diameters 2 cm and 4 cm, respectively. The two smaller pipes rejoin at the end, with the flow again going into a 6 cm inside diameter pipe. The velocity of flow in the 6 cm diameter pipe is 1 m/s, and the velocity in the 2 cm diameter branched pipe is 0.5 m/s. Most nearly, what is the velocity of flow in the branched 4 cm diameter pipe?

(A) 1.2 m/s

(B) 2.1 m/s

(C) 4.3 m/s

(D) 7.0 m/s

**6.** A centrifugal pump has a flow rate of 1 L/s for a rotational speed of 2000 rpm. Assuming a constant impeller diameter, the flow rate for a rotational speed of 3000 rpm is most nearly

(A) 0.75 L/s

(B) 1.5 L/s

(C) 2.0 L/s

(D) 3.0 L/s

**7.** The head generated by a centrifugal pump with a 1 m impeller diameter is 100 m of water at a rotational speed of 1000 rpm. If the pump rotational speed is doubled to 2000 rpm without changing the impeller diameter, the head generated by the pump would be most nearly

(A) 50 m

(B) 150 m

(C) 200 m

(D) 400 m

**8.** A centrifugal pump with a 1 m impeller diameter and a 1000 rpm rotational speed discharges 1 L/s and consumes 10 kW. If the discharge needed is 2 L/s, the power consumption for a similar pump with the same diameter would be most nearly

(A) 20 kW

(B) 40 kW

(C) 80 kW

(D) 95 kW

**9.** A venturi meter needs to be installed in a pipe of inner diameter 8 cm. The venturi meter's throat diameter is 3.5 cm. A mercury manometer is used to measure the pressure drop. If the minimum pressure drop that is practicable is 1 cm, the minimum flow of water at 25°C that can be recorded is mostly nearly

(A) $1.6 \times 10^{-3}$ m³/s

(B) $1.4 \times 10^{-2}$ m³/s

(C) $1.5 \times 10^{-2}$ m³/s

(D) $1.6 \times 10^{-1}$ m³/s

**10.** An orifice meter is placed in a pipe that has a 4 cm internal diameter. The maximum pressure drop that can be measured by the mercury manometer is 10 cm Hg. The diameter of the orifice is 1 cm. The maximum flow rate of water at 20°C that can be recorded is

(A) $1.1 \times 10^{-4}$ m³/s

(B) $1.5 \times 10^{-4}$ m³/s

(C) $1.7 \times 10^{-4}$ m³/s

(D) $2.5 \times 10^{-4}$ m³/s

**11.** An open channel has a width of 5 m and a water height of 2 m. The value of the hydraulic radius is most nearly

(A) 1.1 m

(B) 2.0 m

(C) 2.5 m

(D) 10 m

**12.** An open channel has a cross-sectional area of flow of 0.5 m², a hydraulic radius of 0.15 m, and a roughness coefficient of 0.15. Most nearly, what is the slope of the hydraulic gradient needed to achieve a flow rate of 10 L/s?

(A) $1.1 \times 10^{-4}$

(B) $6.7 \times 10^{-4}$

(C) $1.1 \times 10^{-3}$

(D) $6.7 \times 10^{-3}$

**13.** In a circular sewer pipe of diameter 60 cm, the average velocity of full flow is 1.1 m/s. At a depth of 20 cm, the average velocity of flow would be most nearly

(A) 0.70 m/s

(B) 1.4 m/s

(C) 2.1 m/s

(D) 2.8 m/s

**14.** A circular sewer of diameter 30 cm has sewage flowing at a depth of 10 cm. The area of cross section of the flow is most nearly

(A) 0.010 m²

(B) 0.011 m²

(C) 0.016 m²

(D) 0.020 m²

**15.** Water at 25°C is flowing in a commercial steel pipe of diameter 6 cm. The velocity of flow is 10 m/s, and the head loss is 0.5 m/m of pipe. The loss of power due to friction per meter of pipe is

(A) 0.014 kW/m

(B) 0.14 kW/m

(C) 0.95 kW/m

(D) 0.98 kW/m

**16.** A pitot tube is inserted at the center of a pipe that has a 10 cm diameter. Water at 25°C flows through the pipe and extends up the pitot tube a distance of 1 cm from the outside of the pipe. The pipe has a thickness of 0.5 cm. The velocity of water at the center of the pipe is most nearly

(A) 24 cm/s

(B) 34 cm/s

(C) 54 cm/s

(D) 84 cm/s

**17.** In a packed bed operating at 25°C, a vapor contained in an air stream is dissolved in water. The Henry's law constant of the vapor is 0.10 atm·L/mol. The volumetric flow rate of air is 1 m³/s. Most nearly, what volumetric flow rate of water is required?

(A) 2 L/s

(B) 4 L/s

(C) 6 L/s

(D) 10 L/s

**18.** Primary treatment of wastewater consists of

(A) chlorination

(B) removal of toxic industrial wastes

(C) regulating the flow of the wastewater through tanks of varied sizes so that the flow is a constant in secondary and tertiary treatment

(D) removing solids and particles of various sizes

**19.** The viscosity of water at 25°C is 0.87 cP. For a particle with a diameter of $5 \times 10^{-4}$ m and a density of 1.8 g/cm³, the terminal velocity is most nearly

(A) 0.073 m/s

(B) 0.13 m/s

(C) 1.9 m/s

(D) 2.3 m/s

**20.** A sample of wastewater is diluted by a factor of 1:10. The diluted wastewater has an initial dissolved oxygen concentration of 7.0 mg/L. After 5 d it has a dissolved oxygen concentration of 3.0 mg/L. The 5-day BOD of the initial undiluted wastewater is most nearly

(A) 3 mg/L

(B) 4 mg/L

(C) 7 mg/L

(D) 40 mg/L

**21.** A sample of wastewater has a kinetic rate constant of 0.1/d. The initial dissolved oxygen reading is 8.00 mg/L. The reading after 2 d without any additional oxygen being added is 6.00 mg/L. Therefore, the ultimate BOD is most nearly

(A) 2.0 mg/L

(B) 9.0 mg/L

(C) 11 mg/L

(D) 21 mg/L

**22.** The reason for measuring turbidity of water is

(A) high turbidity requires better filtration operation coupled with coagulation

(B) high turbidity can cause problems with removal of biological oxygen demand (BOD)

(C) high turbidity increases pumping costs

(D) high turbidity means the water must spend more time in the sedimentation chamber

**23.** A sample of water has the following cation concentrations.

| cation | concentration (mg/L) |
|---|---|
| Na | 10 |
| Ca | 20 |
| Mg | 20 |

The total hardness of the water is most nearly

(A) 20 mg/L CaCO₃

(B) 40 mg/L CaCO₃

(C) 130 mg/L CaCO₃

(D) 140 mg/L CaCO₃

**24.** The following data are available for the filtration of wastewater in a treatment plant

superficial velocity of the wastewater: 0.3 m/s

average diameter of the particles in the bed: 0.03 m

porosity of the bed: 47%

depth of the bed: 20 m

The head loss through the bed is most nearly

(A) 0.54 m

(B) 5.6 m

(C) 56 m

(D) 110 m

**25.** The equilibrium constant, $K_{eq}$, is $5 \times 10^{-11}$ mol/L for the reaction

$$HCO_3^- \leftrightarrow H^+ + CO_3^{-2}$$

The molar concentration of $HCO_3^-$ at a pH of 7.5 is most nearly

(A) $3.8 \times 10^{-10}$ mol/L

(B) $4.0 \times 10^{-5}$ mol/L

(C) $2.0 \times 10^{-5}$ mol/L

(D) 1.0 mol/L

**26.** Sludge is digested anaerobically after being removed from a secondary treatment tank. During the digestion process, the gas that is given out is

(A) carbon dioxide

(B) methane

(C) nitrogen oxide

(D) hydrogen sulfide

**27.** The chemical oxygen demand (COD) of octane is the amount of oxygen required to convert the compound to carbon dioxide and water. The molecular formula of octane is $C_8H_{18}$. The COD of octane is most nearly

(A) 3.5 g $O_2$/g octane

(B) 16 g $O_2$/g octane

(C) 23 g $O_2$/g octane

(D) 45 g $O_2$/g octane

**28.** The equation describing the overall conversion of ammonia to nitrate in pure water is as follows

$$NH_3 + 2O_2 \rightarrow HNO_3 + H_2O$$

The mass of oxygen required to complete the nitrification of 100 kg of ammonia is most nearly

(A) 190 kg

(B) 230 kg

(C) 380 kg

(D) 460 kg

**29.** The reaction for a biologically degraded contaminant is first order. The half-life of the contaminant is 3 wk. The reaction rate is most nearly

(A) 0.033/d

(B) 0.041/d

(C) 0.15/d

(D) 0.23/d

**30.** The reaction rate constant, $k$, for the reaeration of a stream depends on temperature, $T$, in degrees Celsius is as follows.

$$k(T) = \left(0.2 \, \frac{1}{d}\right) 1.024^{T-20°C}$$

The reaction-rate constant at 5°C is most nearly

(A) 0.14/d

(B) 0.23/d

(C) 0.29/d

(D) 0.32/d

**31.** What is the general solution to the equation shown, in terms of the constants $\alpha$ and $\beta$?

$$\frac{d^2Y}{dx^2} + 16y = 0$$

(A) $\alpha e^{4x} + \beta e^{-4x}$

(B) $\alpha e^{3x} + \beta e^{-3x}$

(C) $\alpha e^{2x} + \beta e^{-2x}$

(D) $\alpha \cos 4x + \beta \sin 4x$

**32.** A stream of flowing water at 20°C initially has an ultimate BOD in the mixing zone of 10 mg/L. The saturated oxygen concentration is 8.9 mg/L, and the initial dissolved concentration is 8.5 mg/L. The reaeration rate is 2.00/d, the deoxygenation rate constant is 0.1/d, and the velocity of the stream is 0.11 km/min. The concentration of dissolved oxygen in the flowing stream after 160 km is most nearly

(A) 0.050 mg/L

(B) 1.5 mg/L

(C) 2.5 mg/L

(D) 8.4 mg/L

**33.** A one-time discharge of a pollutant to a lake occurs. The concentration of the pollutant as a function of time is given in the following table.

| time (d) | concentration (g/L) |
|---|---|
| 0 | 10 |
| 1 | 3.3 |
| 2 | 2.0 |

If $C$ represents concentration in g/L and $t$ represents time in days, the rate expression for the reaction is

(A) $C = 10 \, \frac{g}{L} - (6.7 \text{ d}^{-1})t$

(B) $C = \left(10 \, \frac{g}{L}\right) e^{-(1.1 \text{ d}^{-1})t}$

(C) $C = \left(\left(0.2 \, \frac{L}{g \cdot d}\right)t + \frac{1}{10 \, \frac{g}{L}}\right)^{-1}$

(D) all of the above rate expressions are appropriate

**34.** An experiment shows that 84% of fluchloralin in water photolyzes after 13 d of exposure to natural sunlight. If this reaction has first-order kinetics, the rate constant is most nearly

(A) 0.065/d
(B) 0.012/d
(C) 0.19/d
(D) 0.14/d

**35.** For a shaker baghouse, the net area of woven fabric required to adequately capture limestone dust in a stream with a 50 m³/min flow rate is most nearly

(A) 0.02 m²
(B) 10 m²
(C) 40 m²
(D) 60 m²

**36.** The nearest receptor to a 70 m stack is at a downwind distance of 250 m from the stack. Wind speed is 2 m/s. At these conditions, the effective stack height is 80 m, the horizontal dispersion parameter is 40 m, the vertical dispersion parameter is 24 m, and the plume reaches its point of maximum rise at a downwind distance of 230 m. A compound is emitted from the stack at a rate of 580 μg/s. The concentration of this compound at ground level when the wind is blowing directly from the stack to the nearest receptor is most nearly

(A) $1.3 \times 10^{-4}$ μg/m³
(B) $2.6 \times 10^{-4}$ μg/m³
(C) $3.7 \times 10^{-4}$ μg/m³
(D) $4.4 \times 10^{-4}$ μg/m³

**37.** An expression relating $X_z$, the atmospheric concentration of a compound at altitude $z$, and $X_0$, the atmospheric concentration of the compound at sea level, is

$$X_z = X_0 e^{-\frac{z}{8.400 \text{ km}}}$$

The density of air at sea level and at 20°C is 1.2 kg/m³, and air comprises 21% oxygen by weight. The concentration of oxygen in the air at an altitude of 2 km above sea level is most nearly

(A) 170 g/m³
(B) 200 g/m³
(C) 250 g/m³
(D) 260 g/m³

**38.** The molecular weight of air is 29 g/mol. At any given moment, the flow of carbon dioxide into the atmosphere, $F_{in}$, and the flow of carbon dioxide out of the atmosphere, $F_{out}$, nearly equal each other and can be related as follows.

$$F_{in} = F_{out} = \frac{M}{\tau}$$

In this equation, $M$ represents the total mass of carbon dioxide in the atmosphere and $\tau$ represents the residence time of carbon dioxide in the atmosphere. The total mass of air in the atmosphere is approximately $5.2 \times 10^{18}$ kg, the concentration of carbon dioxide in the atmosphere is 360 ppm, and the residence time of carbon dioxide in the atmosphere is assumed to be 100 yr. The flux of carbon dioxide into and out of the atmosphere is most nearly

(A) $3.4 \times 10^{10}$ kg/d
(B) $5.1 \times 10^{10}$ kg/d
(C) $7.7 \times 10^{10}$ kg/d
(D) $1.4 \times 10^{14}$ kg/d

**39.** A plume is emitted from a 20 m high stack. Wind speed is constant at 3 m/s, there is no wind shear, and the topography is flat. The plume's point of maximum rise occurs at a downwind distance of 600 m from the stack. The emissions from the stack have a buoyancy flux of 55 m⁴/s³. Prior to the point of maximum plume rise, the vertical distance between the top of the stack and the centerline of the plume can be modeled as

$$\Delta h = \frac{1.6 F^{1/3} x^{2/3}}{\mu}$$

The height of the plume's centerline above the ground at a distance of 0.4 km downwind of the stack is most nearly

(A) 110 m
(B) 130 m
(C) 140 m
(D) 160 m

**40.** An air stripper is used to remove a solute from air. The incoming air has a solute concentration of 5 ppm. This air is at 1 atm and 25°C. The temperature change of air during the process can be considered negligible. The outgoing concentration of solute in water is 0.5 ppm, and the incoming concentration of solute in water is 100 ppm. The dimensionless Henry's law constant, $H'$, is 10. The average airflow rate is 1.00 m³/10 s, while the average water flow rate is 0.1 m³/10 s. If the height of transfer unit is 3 m, the height of the column is most nearly

(A) 5.0 m

(B) 16 m

(C) 25 m

(D) 75 m

**41.** A cyclone collects particles that are 20 μm in diameter with 50% efficiency. It has been suggested that the efficiency of collecting particles of this size be increased by increasing the inlet velocity of the particulate-laden gas stream. The current inlet velocity is 6 m/s. In order to increase the cyclone's efficiency for collecting 20 μm particles to 80%, the inlet velocity must be most nearly

(A) 1.5 m/s

(B) 24 m/s

(C) 36 m/s

(D) 420 m/s

**42.** A quarterly leak detection and repair program with a leak definition of 500 ppm to reduce fugitive emissions at a refinery is expected to have an annualized cost of $290,000. The current quarterly leak detection and repair program, which has a 10,000 ppm leak definition, has an annualized cost of $204,000. The lower leak definition of 500 ppm will reduce fugitive emissions by 195 Mg/yr. The value of the product lost in fugitive emissions is $4.20/L. Assume the density of the captured losses is 14 kg/L. The net annualized cost of lowering the leak definition from 10,000 ppm to 500 ppm is most nearly

(A) $27,500/yr

(B) $50,300/yr

(C) $75,500/yr

(D) $136,000/yr

**43.** Values for the concentration of a chemical at a ground-level location 500 m downwind of a stack are given in the following table.

| atmospheric stability class | chemical concentration (g/m³) |
|---|---|
| A | $1.2 \times 10^{-4}$ |
| B | $2.0 \times 10^{-4}$ |
| C | $1.6 \times 10^{-4}$ |
| D | $1.5 \times 10^{-5}$ |
| E | $1.7 \times 10^{-8}$ |
| F | $6.0 \times 10^{-17}$ |

On a summer day with the sun high in the sky, a few broken clouds, and a 4 m/s surface wind, the concentration of the compound at this location is most nearly

(A) $1.7 \times 10^{-8}$ g/m³

(B) $1.5 \times 10^{-5}$ g/m³

(C) $1.2 \times 10^{-4}$ g/m³

(D) $1.8 \times 10^{-4}$ g/m³

**44.** The overall equation for the oxidation of carbon monoxide in the troposphere is

$$CO + 2O_2 + h\nu \rightarrow CO_2 + O_3$$

The overall effect of oxidizing 50 ppb of carbon monoxide is most nearly

(A) a 150 ppb increase in the ozone concentration

(B) a 50 ppb decrease in the oxygen concentration

(C) a 50 ppb increase in the ozone concentration

(D) an 80 ppb increase in the carbon dioxide concentration

**45.** A waste is determined to be characteristically toxic under the Resource Conservation and Recovery Act (RCRA) if

(A) the waste contains specific compounds at greater than threshold concentrations

(B) it is determined that chemicals in the waste are bioaccumulative, toxic, and persistent in the environment

(C) leachate from the waste contains specific compounds at greater than threshold concentrations

(D) exposure to the waste causes adverse health effects in laboratory animals

**46.** Under Superfund legislation, a company can be held liable for all costs associated with remediation of a contaminated site

(A) only if the company disposed of wastes at the site illegally

(B) only if the company is solely responsible for contamination at the site

(C) only if the company disposed of wastes at the site illegally and is solely responsible for contamination at the site

(D) if the company disposed of wastes at the site legally and contributed only a small fraction of wastes disposed of at the site

**47.** A city has municipal solid waste with the following characteristics.

| waste component | mass (%) (dry basis) |
|---|---|
| food | 3.2 |
| glass | 0.2 |
| metal | 1.2 |
| plastics | 18.9 |
| wood debris | 3.8 |
| paper | 15.8 |
| yard waste | 33.0 |

The mass percent of water in the solid waste is most nearly

(A) 12.0

(B) 23.9

(C) 35.9

(D) 47.8

**48.** A hazardous waste incinerator is required to have a destruction and removal efficiency of 99.99% for tetrachloromethane. The mass flow rate of tetrachloromethane into the incinerator, $W_{in}$, is $1.1 \times 10^{-3}$ kg/s. The maximum allowable mass flow rate of tetrachloromethane out of this incinerator, $W_{out}$, is most nearly

(A) $1.1 \times 10^{-7}$ kg/s

(B) $1.1 \times 10^{-6}$ kg/s

(C) $9.9 \times 10^{-5}$ kg/s

(D) $9.9 \times 10^{-4}$ kg/s

**49.** The most difficult compound to incinerate in a stream is the one with the highest incinerability index, $I$. The formula is

$$I = C + \frac{100 \frac{\text{kcal}}{\text{g}}}{H}$$

$C$ is mass percent and $H$ is heating value. Compound concentration and heating values for a stream are

| compound | $C$ (mass %) | $H$ (kcal/g) |
|---|---|---|
| acetonitrile | 3.2 | 7.37 |
| benzene | 5.7 | 10.03 |
| naphthalene | 4.7 | 9.62 |
| vinyl chloride | 3.0 | 4.45 |

The most difficult compound to incinerate in this stream is

(A) acetonitrile

(B) benzene

(C) naphthalene

(D) vinyl chloride

**50.** The water level of a well drilled into an artesian aquifer is

(A) at the same level as the top of the water in the aquifer

(B) above the level of the top of the water in the aquifer

(C) below the level of the top of the water in the aquifer

(D) either at the same level as or above the level of the top of the water in the aquifer

**51.** Assume a first-order reaction for destruction of chloroform. For chloroform, the frequency factor is $2.90 \times 10^{12}$/s and the activation energy is 49 kcal/mol. The temperature needed to achieve 99.99% destruction of chloroform in an incinerator treating contaminated soil that has a 1.2 s residence time is most nearly

(A) 800K

(B) 930K

(C) 1000K

(D) 2000K

**52.** A flow pattern is a three-dimensional representation of total head. When water is pumped out of an unconfined aquifer, the flow pattern of the groundwater has

(A) no appreciable change due to removal of water at the well

(B) a linear slope of the total head of water flowing to the well

(C) a cone of depression with the tip of the cone being at the center point of the well

(D) a semi-spherical depression

**53.** Some soil has a discharge rate of 0.1 L/s, an area of 11 m², and a hydraulic head that is given empirically by the function

$$h = -0.03x + 0.3$$

The units of $h$ and $x$ in this equation are in meters. The hydraulic conductivity of this soil is most nearly

(A) $3 \times 10^{-4}$ m/s

(B) $3 \times 10^{-3}$ m/s

(C) $3 \times 10^{-2}$ m/s

(D) $3 \times 10^{-1}$ m/s

**54.** Which of the following is true of the meanings of accuracy and precision?

(A) Both the precision and the accuracy of an instrument reflect the number of significant digits in an instrument reading.

(B) Both the precision and the accuracy of an instrument reflect how close an instrument reading is to the true value of what is being measured.

(C) The precision of an instrument reflects the number of significant digits in an instrument reading, while the accuracy of an instrument reflects how close an instrument reading is to the true value of what is being measured.

(D) The precision of an instrument reflects how close an instrument reading is to the true value of what is being measured, while the accuracy of an instrument reflects the number of scientific digits in an instrument reading.

**55.** What is the first part of the solution for the differential equation shown?

$$\frac{d^2Y}{dx^2} + Y = 4$$

(A) $Y = 2X^2 +$ other terms that are functions of $X$

(B) $Y = -2X^2 +$ other terms that are functions of $X$

(C) $Y = A\cos + \beta +$ other terms that are functions of $X$

(D) $Y = A\cos + \beta\sin +$ other terms that are functions of $X$

**56.** The rate of dry deposition of particles from the atmosphere in still air can be modeled using Stokes' law. Compare the rate of deposition of particles that are 15 μm in diameter to that of particles that are 10 μm in diameter, assuming that all particles have the same density. The deposition rate of a particle 15 μm is most nearly

(A) 0.44 times the deposition rate of a particle 10 μm in diameter

(B) 1.5 times the deposition rate of a particle 10 μm in diameter

(C) 2.3 times the deposition rate of a particle 10 μm in diameter

(D) 4.6 times the deposition rate of a particle 10 μm in diameter

**57.** At 15°C, the Henry's law constant for ammonia is 0.62 atm. The concentration of ammonia in a water solution at this temperature is $8.1 \times 10^{-3}$ mol/L. The partial pressure of ammonia vapor above the liquid phase is most nearly

(A) $9.0 \times 10^{-5}$ atm

(B) $4.4 \times 10^{-5}$ atm

(C) $3.8 \times 10^{-5}$ atm

(D) $5.0 \times 10^{-3}$ atm

**58.** Hair conditioner is produced at a factory in a batch reactor. Three batches of conditioner are made every day, seven days a week. After each batch is complete, the conditioner is drained out of the reactor through a hose to containers that carry it to the factory's bottling machines. Approximately 2.6 L/batch of product clings to the inside of the reaction vessel and the transfer hose instead of draining to the container that carries product to the bottling machines. This lost product, called cling loss, goes to the facility's wastewater treatment plant when the reactor and the hose are rinsed out with hot water in preparation for the next batch. The conditioner contains emulsified oils, and each liter of product sent to the wastewater treatment plant results in the generation of 2 L of sludge solids at the wastewater treatment plant. These sludge solids must be trucked to a landfill at a cost of $1.50/L.

The factory is considering the use of mechanical scrapers, including squeegees to scrape the sides of the reaction vessel and pigs to pass through the transfer hose, to reduce cling losses. Use of the mechanical scrapers would reduce the load on the wastewater treatment plant and make more conditioner available for bottling. Properly used, the mechanical scrapers would not only reduce the volume of cling losses to 0.8 L/batch, but would also reduce the amount of rinsing required by a third. The factory decides to switch to using squeegees and a pig before rinsing out the reactor and the transfer hose.

The landfill costs saved by the factory every year are most nearly

(A) $3100/yr
(B) $3800/yr
(C) $4400/yr
(D) $5900/yr

**59.** A venturi meter with a throat diameter of 2 in is placed in a 5 in horizontal pipe carrying water. The pressure drop in the meter is 6 mm Hg. The volumetric flow rate through the pipe is most nearly

(A) $0.0011 \text{ m}^3/\text{s}$
(B) $0.0022 \text{ m}^3/\text{s}$
(C) $0.0027 \text{ m}^3/\text{s}$
(D) $0.0036 \text{ m}^3/\text{s}$

**60.** Arsenic has a $5 \times 10^{-5}$ $(\mu g/L)^{-1}$ drinking water unit risk. The excess cancer cases due to arsenic in drinking water predicted for one million people drinking water with an arsenic concentration of 2 ppb all of their lives is most nearly

(A) 25 cases
(B) 100 cases
(C) 40,000 cases
(D) 500,000 cases

**61.** A person 10 m from a source of gamma or x-rays will receive a dose that is most nearly which fraction of the dose they would receive at a distance of 1 m from the same source?

(A) 0.005
(B) 0.05
(C) 0.01
(D) 0.1

**62.** The half-life of $^{68}_{31}\text{Ga}$ is 68.3 min. The percentage of $^{68}_{31}\text{Ga}$ remaining at the end of 6 hr is most nearly

(A) 2.1%
(B) 2.6%
(C) 8.9%
(D) 38%

**63.** An idealized compression process is proposed for a new device. The pressure applied, in units of atm, is proportional to 1.1 times the volume, in units of m³. If the process compresses material from 0.8 m³ to 0.6 m³, the work done on the material is most nearly

(A) $0.104 \text{ atm·m}^3$
(B) $0.154 \text{ atm·m}^3$
(C) $0.204 \text{ atm·m}^3$
(D) $0.304 \text{ atm·m}^3$

**64.** The partial pressure of carbon monoxide (CO) in a given room is 1 mm Hg. The temperature in the room is 29°C, and the pressure is 1 atm. The capacity of the room is 2000 L. The mass of CO in the room is most nearly

(A) 1.6 g
(B) 3.0 g
(C) 3.1 g
(D) 3.5 g

**65.** A project in a plant has an initial expense of $15,000,000, and will cost an additional $1,000,000 in expenses each year for 15 years. Completing the project will produce an additional $3,000,000 in revenue each year for 15 years, and at the end of 15 years, will produce $15,000,000 in additional income. The interest rate is 5%. The net future worth of the project at the end of 15 years is most nearly

(A) $16,000,000
(B) $22,000,000
(C) $27,000,000
(D) $31,000,000

**66.** A project plan has an expected initial investment of $100,000,000 and expenses of $1,000,000 a year for 20 years. The minimum expected income each year is $15,000,000 for 20 years. The interest rate is 5%. The benefit-cost ratio based on the present worth is most nearly

(A) 1.3
(B) 1.7
(C) 2.5
(D) 3.0

**67.** A gas mixture is at a pressure of 1 atm. By volume, the mixture contains 60% nitrogen ($N_2$), 30% oxygen ($O_2$), 5% carbon monoxide (CO), and 5% carbon dioxide ($CO_2$). The mixture is at a pressure of 1 atm. The mass percentage of $CO_2$ in the mixture is most nearly

(A) 5.3%
(B) 6.3%
(C) 7.3%
(D) 8.3%

**68.** The solubility product for magnesium hydroxide is $2 \times 10^{-11}$ mol$^3$/L$^3$. The pH of a saturated solution of magnesium hydroxide is most nearly

(A) 4.5
(B) 8.7
(C) 10.4
(D) 10.5

**69.** What is the degree of the carbon labeled 4 in the molecule shown?

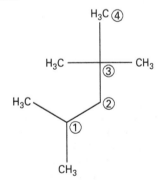

(A) 1
(B) 2
(C) 3
(D) 4

**70.** During halogenation of an alkane, a hydrogen atom is replaced with a halogen. During halogenation of the molecule shown, which carbon atom is most likely to have an attached hydrogen replaced with a halogen?

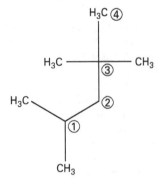

(A) 1
(B) 2
(C) 3
(D) 4

**71.** According to International Union of Pure and Applied Chemistry nomenclature, what is name of the molecule shown?

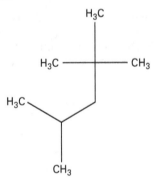

(A) 1-dimethyl-3-dimethylbutane
(B) 2-methyl-4-dimethylpentane
(C) 2,2,4-trimethylpentane
(D) 1,1,3,3-tetramethylbutane

**72.** The difference in mass between 250 mL of water at 6°C and 250 mL of water at 45°C is most nearly

(A) 0
(B) 1.4 g
(C) 2.5 g
(D) 3.0 g

**73.** 0.02 mg of chlorpyrifos is added to 103 L of a mixture of water and octanol. The water phase has a volume of 100 L and the octanol phase has a volume of 3 L. The octanol-water partition coefficient for chlorpyrifos is

4.7. Once equilibrium is established, the concentration of chlorpyrifos in the octanol phase is most nearly

(A) $8.2 \times 10^{-4}$ mg/L

(B) $8.5 \times 10^{-4}$ mg/L

(C) $9.1 \times 10^{-4}$ mg/L

(D) $4.3 \times 10^{-3}$ mg/L

**74.** A study estimates that the average non-occupational human ingestion of methyl mercury in the United States is approximately 1 $\mu$g/kg·d. The reference dose for methyl mercury is 0.0001 mg/kg·d. The noncarcinogenic hazard index due to ingesting the average non-occupational amount of methyl mercury is most nearly

(A) 0.0001

(B) 0.1

(C) 10

(D) 10,000

**75.** A watershed has 2 ac of woods upstream of 18 ac of concrete parking lot. The concrete pavement has a runoff coefficient of 0.90 for 2- to 10-year storms, and the woods have a runoff coefficient of 0.05 for 2- to 10-year storms. The overall runoff coefficient for this watershed for a 2- to 10-year storm is most nearly

(A) 0.14

(B) 0.82

(C) 0.86

(D) 0.90

**76.** The table shown provides inflows and outflows for a water supply reservoir.

| period | fall | winter | spring | summer |
|---|---|---|---|---|
| inflow (km³) | 1 | 7 | 1 | 0 |
| demand (km³) | 2 | 0 | 2 | 5 |

Assume no variation from year to year, and neglect seepage and evaporation losses. The active reservoir capacity needed in order to just meet demand is most nearly

(A) 5 km³

(B) 6 km³

(C) 7 km³

(D) 9 km³

**77.** A planned development is anticipated to cause a 15-fold increase in the phosphorus content in runoff from the area. The developer must construct stormwater management strategies to remove the additional phosphorus before stormwater is discharged to surface waters outside of the development area. The phosphorus removal efficiency required is most nearly

(A) 6.7%

(B) 15%

(C) 85%

(D) 93%

**78.** A shallow well in a confined aquifer has a transmissivity of 1300 ft²/day. The aquifer is 45 ft deep. The hydraulic conductivity of this well is most nearly

(A) 0.035 ft/day

(B) 29 ft/day

(C) 45 ft/day

(D) 1300 ft/day

**79.** In order to apply Jacob's method for calculating transmissivity of a nonleaky confined aquifer, the incremental change in drawdown for one log cycle of time must be determined. A plot of the drawdown for a particular aquifer as a function of the base 10 logarithm of time is shown.

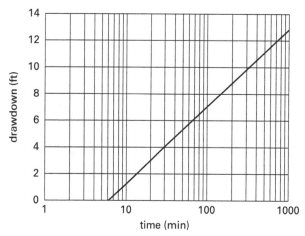

The incremental change in drawdown for one log cycle of time is most nearly

(A) 0.012 ft

(B) 3.0 ft

(C) 4.3 ft

(D) 5.8 ft

**80.** An aquifer has a hydraulic conductivity of 67 m/d, a hydraulic gradient of 0.5%, and an effective porosity of 15%. The Darcy velocity is most nearly

(A) 0.33 m/d

(B) 10 m/d

(C) 33 m/d

(D) 220 m/d

**81.** The dimensions and flow rate of a constant head permeameter are shown. The permeameter is cylindrical, with a diameter of 0.37 m. The permeameter discharges to the atmosphere.

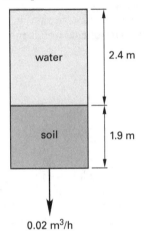

The hydraulic conductivity of the soil sample is most nearly

(A) 0.02 m/h

(B) 0.06 m/h

(C) 0.08 m/h

(D) 0.14 m/h

**82.** Soil in an unsaturated zone is a combination of gaseous, solid, and aqueous phases. The soil has been contaminated with a non-aqueous phase liquid (NAPL). The contaminant may be found in

(A) the solid phase portion of the soil only

(B) the solid and gaseous phase portions of the soil only

(C) the solid, gaseous, and aqueous phase portions of the soil

(D) a NAPL phase plus the solid, gaseous, and aqueous phase portions of the soil

**83.** A series of lab tests concludes that a given dosage of chlorine added to a water sample requires 1 min to bring the concentration of microorganisms down to 50% of the initial concentration. The time needed to bring the concentration of microorganisms in the sample down to 0.1% of the initial concentration is most nearly

(A) 4 min

(B) 6 min

(C) 10 min

(D) 15 min

**84.** The chlorination in a baffled chamber needs a CT factor of 12 in order to make safe drinking water. The average chlorine concentration is 20 mg/L, and the peak chlorine concentration is 1.8 times the average concentration. Baffling conditions are average. The time of contact for the water with the disinfectant is most nearly

(A) 0.50 min

(B) 0.67 min

(C) 1.00 min

(D) 1.33 min

**85.** The average water consumption rate for a small city is 175 gpcd, and is expected to rise at the rate of 0.1% per year for the foreseeable future due to new construction projects. The population is expected to have a geometric growth rate of 0.1 per year. After 10 years, the percentage increase in total water usage is most nearly

(A) 10%

(B) 80%

(C) 100%

(D) 170%

**86.** The per capita water usage in a city with a population of 540,000 is 181 gpcd. The population is expected to increase at a rate of 1.1% per year due to internal migration, and the per capita water usage is expected to increase at a rate of 0.02% per year. The expected total water usage in the city 10 years from now is most nearly

(A) 100,000,000 gpd

(B) 110,000,000 gpd

(C) 120,000,000 gpd

(D) 130,000,000 gpd

**87.** Water from one tank is being discharged to another tank. The difference in the water levels between the two tanks is 25 m. A pipe 1000 m in length connects the two tanks. The first 500 m of the pipe has a diameter of 50 cm, and the last 500 m of pipe has a diameter of

30 cm. The coefficient of friction is 0.0055. The total flow rate through the pipe is most nearly

(A) 1.00 m³/s

(B) 1.40 m³/s

(C) 1.60 m³/s

(D) 1.80 m³/s

**88.** Two tanks containing water are connected by two pipes in parallel. The difference in the elevation of the water levels is 4 m. The first pipe is 2500 m long with a diameter of 1 m and a Darcy coefficient of 0.1; the second pipe is 2498 m long with a diameter of 1.1 m and Darcy coefficient of 0.072. The total flow between the tanks is most nearly

(A) 0.89 m³/s

(B) 1.1 m³/s

(C) 2.0 m³/s

(D) 2.2 m³/s

**89.** A 10 m long rectangular weir has a water head of 50 cm. The end contractions of the weir have been suppressed. The flow rate over the weir is most nearly

(A) 42 m³/min

(B) 140 m³/min

(C) 210 m³/min

(D) 390 m³/min

**90.** Two tanks of equal volume are interconnected, with a valve between them. One tank contains nitrogen gas at 20°C and 1.2 atm, and the other contains nitrogen gas at 25°C and 0.8 atm. The valve is closed for a long time, then at time $t = 0$, the valve is opened. Assume both tanks are perfectly insulated. The final temperature of the contents of both tanks is most nearly

(A) 22°C

(B) 22.5°C

(C) 23°C

(D) 23.5°C

**91.** The specific heat of a gas at a constant, low pressure is given by the equation shown.

$$c_p = (0.9 + 0.8\,T)R$$

The equation for the specific heat of the gas at constant volume is

(A) $c_v = 0.8RT + 0.1R$

(B) $c_v = 0.8RT$

(C) $c_v = 0.1R$

(D) $c_v = 0.8RT - 0.1R$

**92.** Which statement about life-cycle cost analysis of a transportation project is true?

(A) It is used to evaluate the cost-effectiveness of developing, operating, and salvaging or disposing of competing alternatives.

(B) It applies a discount rate to determine the present value of future cash flows.

(C) It might include user costs.

(D) all of the above

**93.** The molecular formula for vinyl chloride is

(A) $CH_2{=}CHCl$

(B) $CHCl{=}CHCl$

(C) $CH_3CH_2Cl$

(D) $COCl_2$

**94.** The common name of the molecule shown is

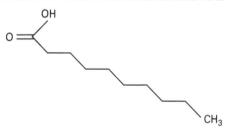

(A) capric acid

(B) pelargonaldehyde

(C) caprylone

(D) 1-methoxynonane

**95.** During the Gold King Mine incident, which began on August 5, 2015, many heavy metals, including arsenic, were released from mine tailing leachate into the Animas River near Silverton, Colorado. Farmington, New Mexico is a downstream community which draws water from three rivers, as shown: the Animas, the San Juan, and the La Plata. On August 8, 2015, arsenic was present at point A at a concentration of 11 μg/L. On this day, the average flow rate at point B was 1673 ft³/

sec, the average flow rate at point C was 776 ft³/sec, and the average flow rate at point D was 2 ft³/sec.

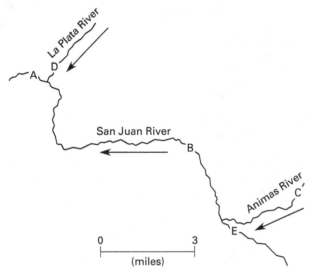

Assume the concentration of arsenic at points D and E was negligible, and that any drop in concentration between points C and A was due to dilution. The concentration of arsenic in the Animas River before it met the San Juan River was most nearly

(A) 5 μg/L

(B) 13 μg/L

(C) 23 μg/L

(D) 35 μg/L

**96.** Acute toxicity testing of oral ingestion of thymoquinone in rats produced the results shown.

| number of rats dosed | dose (mg/kg of body weight) | number of rats that died |
|---|---|---|
| 10 | 100 | 0 |
| 10 | 500 | 1 |
| 10 | 1000 | 5 |
| 10 | 1500 | 7 |
| 10 | 2000 | 10 |

According to these results, the oral $LD_{50}$ of thymoquinone for rats is most nearly

(A) 500 mg/kg of body weight

(B) 1000 mg/kg of body weight

(C) 1500 mg/kg of body weight

(D) 2000 mg/kg of body weight

**97.** A workroom at 25°C containing open containers of toluene is unvented and is saturated with toluene. The vapor pressure of toluene at this temperature is 3.8 kPa. The permissible level of workplace inhalation exposure to toluene for an 8 hr period is a time-weighted average of 200 ppm, according to the U.S. Occupational Safety and Health Act. The concentration of toluene in the workroom is most nearly

(A) 3.8% of the 8 hr time-averaged permissible exposure level

(B) 13% of the 8 hr time-averaged permissible exposure level

(C) 19,000% of the 8-hr time-averaged permissible exposure level

(D) 76,000% of the 8 hr time-averaged permissible exposure level

**98.** A chemical's reference dose

(A) decreases as the chemical's noncarcinogenic toxicity increases

(B) has no relationship with a chemical's toxicity

(C) is a metric that expresses the potency of a chemical's carcinogenic effects

(D) increases as the chemical's noncarcinogenic toxicity increases

**99.** The median concentration of manganese in private wells in the U.S. is 0.007 mg/L. The lifetime ingestion of manganese by a typical adult woman drinking well water containing the median concentration of manganese every day for her entire lifetime is most nearly

(A) 200 mg/lifetime

(B) 380 mg/lifetime

(C) 440 mg/lifetime

(D) 13,000 mg/lifetime

**100.** An area of grassland previously used for grazing is developed into townhouses. The grassland is hydrologic soil group type C, and is in good condition before development. Assuming median moisture conditions, the curve number of this area is 74 before development and 90 after development. The initial abstraction is equivalent to 0.2 times the maximum basin retention both before and after development. The increase in runoff from an isolated 1 in rainstorm over the area is a factor of most nearly

(A) 1.2

(B) 5.3

(C) 13

(D) 14

Table for Prob. 101

| period | Jan | Feb | Mar | Apr | May | Jun | Jul | Aug | Sep | Oct | Nov | Dec |
|---|---|---|---|---|---|---|---|---|---|---|---|---|
| inflow (km³) | 1 | 2 | 4 | 7 | 11 | 6 | 2 | 2 | 1 | 0 | 0 | 0 |
| demand (km³) | 5 | 3 | 1 | 0 | 0 | 0 | 1 | 2 | 4 | 5 | 9 | 6 |

**101.** *Table for Prob. 101* provides inflows and outflows for a water supply reservoir. Assume there is no variation from year to year.

Neglect seepage and evaporation losses. The active reservoir capacity needed in order to just meet demand is most nearly

(A) 9 km³

(B) 23 km³

(C) 28 km³

(D) 36 km³

**102.** A flood control reservoir is dry, with no inflow or outflow, until after a storm begins. The inflow and outflow volumes are shown in the figure.

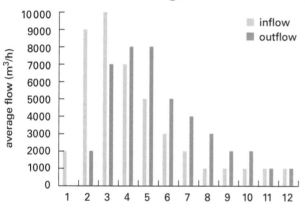

Assume there is no seepage or evaporation. The peak water held in the reservoir during this storm event is most nearly

(A) 7000 m³

(B) 10,000 m³

(C) 12,000 m³

(D) 43,000 m³

**103.** A secondary clarifier is located downstream of an aeration tank, as shown. The solids concentration in the clarified stream from the secondary clarifier is 10% of the suspended solids concentration in the aeration tank. The recycle ratio is 0.25.

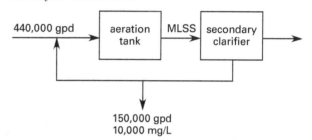

The effluent suspended solids concentration is most nearly

(A) 240 mg/L

(B) 500 mg/L

(C) 530 mg/L

(D) 860 mg/L

**104.** The concentration of suspended solids in an aeration tank is 2500 mg/L. The flow rate from a primary clarifier into the aeration tank is 3,600,000 gpd. The concentration of suspended solids in this flow rate is 2000 mg/L. The aeration tank is also fed with returned activated sludge with a concentration of suspended solids of 9000 mg/L. The flow rate of the returned sludge is most nearly

(A) 200,000 gpd

(B) 260,000 gpd

(C) 280,000 gpd

(D) 1,400,000 gpd

**105.** A septic system has 200 ft of conventional absorption trenches with perforated distribution pipe. The inner diameter of the pipe is 3.0 in. The pipe volume is most nearly

(A) 9.8 gal

(B) 73 gal

(C) 290 gal

(D) 1400 gal

**106.** A non-recirculating municipal wastewater treatment biotower containing modular plastic media has a 5 m diameter. The influent BOD is 180 mg/L. An effluent BOD of 20 mg/L is desired. The average wastewater flow is 450 m³/d. The biotower is operated at 15°C. The required height of the biotower media is most nearly

(A) 0.14 m
(B) 1.1 m
(C) 4.6 m
(D) 5.5 m

**107.** A falling-head permeameter has a soil sample depth of 0.27 m. The permeameter's reservoir tube has the same cross-sectional area as the soil sample. During the test, the water level drops from 0.41 m to 0.023 m in 2 h and 37 min. The hydraulic conductivity of the soil is most nearly

(A) $8.3 \times 10^{-5}$ m/s
(B) $9.1 \times 10^{-5}$ m/s
(C) $1.1 \times 10^{-4}$ m/s
(D) $5.0 \times 10^{-3}$ m/s

**108.** The upper boundary of a water table is defined as the point where the soil becomes saturated with water. The scenario that describes soil that is saturated with water is

(A) water volume/void volume = 1
(B) void volume/total volume = 0
(C) weight of water + weight of soil solids = total weight
(D) weight of water + weight of air = weight of water

**109.** A blower is required to compress 3000 ft³/min from a pressure of 1 atm to a pressure of 2.5 atm at a temperature of 25°C. The efficiency of the blower is 68%. The expected power of the compressor is most nearly

(A) 100 hp
(B) 200 hp
(C) 300 hp
(D) 400 hp

## SOLUTIONS

**1.** The ratio of peak to average daily flow rate and minimum to average daily flow rate can be derived from a sewage flow ratio curves graph. [**Sewage Flow Ratio Curves**]

Reading from the curves, for a population of 30,000,

$$\frac{\text{minimum daily flow rate}}{\text{average daily flow rate}} = 0.4$$

$$\frac{\text{peak daily flow rate}}{\text{average daily flow rate}} = 2.5$$

Therefore, the ratio of the minimum daily flow rate to the peak daily flow rate is

$$\frac{\text{minimum daily flow rate}}{\text{peak daily flow rate}} = \frac{0.4}{2.5}$$
$$= 0.16$$

**The answer is (A).**

**2.** The population is increasing linearly and is expected to be 1.4 million in 2010. The per-capita water consumption rate can be assumed to be increasing linearly to 0.9 kg/person/d in 2010. Therefore,

total consumption in 2010 = (per-capita consumption)
$$\times \text{(population)}$$
$$= \left(0.9 \; \frac{\frac{\text{kg}}{\text{person}}}{\text{d}}\right)$$
$$\times (1.4 \times 10^6 \text{ people})$$
$$= 1.3 \times 10^6 \text{ kg/d}$$

**The answer is (C).**

**3.** Dumping treated sewage discharge into oceans is allowed at present. However, if the substance is dumped directly, it can rise to the surface with the possibility of polluting the coast. Multiport diffusers maximize the amount of diffusion and thus maximize the dispersion and natural cleaning process.

**The answer is (C).**

**4.** The runoff coefficient is given as 0.25, the rainfall intensity is 2.5 cm/h, and the watershed area is 50 000 m². The rational formula for discharge is

$$Q = CIA$$
$$= (0.25)\left(2.5 \ \frac{\text{cm}}{\text{h}}\right)(50\,000 \ \text{m}^2)\left(\frac{1 \ \text{m}}{100 \ \text{cm}}\right)\left(\frac{1 \ \text{h}}{3600 \ \text{s}}\right)$$
$$= 0.087 \ \text{m}^3/\text{s}$$

**The answer is (B).**

**5.** A material balance equation can be used to solve this problem. Let A denote the inlet 6 cm pipe, B denote the 2 cm pipe, and C denote the 4 cm pipe. The incoming flow rate is equal to the sum of the flow rates in the two branch pipes.

$$Q_A = Q_B + Q_C$$
$$\frac{v_A \pi D_A^2}{4} = \frac{v_B \pi D_B^2}{4} + \frac{v_C \pi D_C^2}{4}$$

Solving for $v_C$ gives

$$v_C = \frac{v_A D_A^2 - v_B D_B^2}{D_C^2}$$
$$= \frac{\left(1 \ \frac{\text{m}}{\text{s}}\right)(0.06 \ \text{m})^2 - \left(0.5 \ \frac{\text{m}}{\text{s}}\right)(0.02 \ \text{m})^2}{(0.04 \ \text{m})^2}$$
$$= 2.1 \ \text{m/s}$$

**The answer is (B).**

**6.** The relationship between flow rate and pump rotational speed at a constant impeller diameter is found from the scaling law formula.

*Fans, Pumps, and Compressors*

$$\left(\frac{Q}{ND^3}\right)_2 = \left(\frac{Q}{ND^3}\right)_1$$
$$\frac{Q_2}{Q_1} = \frac{(ND^3)_2}{(ND^3)_1}$$
$$\frac{Q_2}{Q_1} = \frac{N_2}{N_1}$$

In this equation, $Q$ is flow rate and $N$ is rotational speed. Solving for $Q_2$ and substituting the given values yields

$$Q_2 = \frac{Q_1 N_2}{N_1}$$
$$= \frac{\left(1 \ \frac{\text{L}}{\text{s}}\right)\left(3000 \ \frac{\text{rev}}{\text{min}}\right)}{2000 \ \frac{\text{rev}}{\text{min}}}$$
$$= 1.5 \ \text{L/s}$$

**The answer is (B).**

**7.** The formula relating head generated to pump rotational speed for a pump of constant impeller diameter is found from the scaling law formula.

*Fans, Pumps, and Compressors*

$$\left(\frac{H}{N^2 D^2}\right)_2 = \left(\frac{H}{N^2 D^2}\right)_1$$
$$\frac{H_2}{H_1} = \frac{(N^2 D^2)_2}{(N^2 D^2)_1}$$

The diameter is constant, so the head is proportional to the square of the speed. This simplifies to

$$\frac{H_2}{H_1} = \frac{N_2^2}{N_1^2}$$

In this equation, $H$ is head and $N$ is rotational speed. Solving for $H_2$ and substituting the given values yields

$$H_2 = H_1 \left(\frac{N_2}{N_1}\right)^2$$
$$= \frac{(100 \ \text{m})\left(2000 \ \frac{\text{rev}}{\text{min}}\right)^2}{\left(1000 \ \frac{\text{rev}}{\text{min}}\right)^2}$$
$$= 400 \ \text{m}$$

**The answer is (D).**

**8.** The relationship between the power consumption and the discharge, $Q$, for a centrifugal pump of constant impeller diameter can be represented using the scaling laws.

*Fans, Pumps, and Compressors*

$$\left(\frac{Q}{ND^3}\right)_2 = \left(\frac{Q}{ND^3}\right)_1$$

Since the diameter is the same in both cases,

$$\left(\frac{Q}{N}\right)_2 = \left(\frac{Q}{N}\right)_1$$

The scaling law for the power is

**Fans, Pumps, and Compressors**

$$\left(\frac{\dot{W}}{\rho N^3 D^5}\right)_2 = \left(\frac{\dot{W}}{\rho N^3 D^5}\right)_1$$

From the simplified equation for the relationship between the power consumption and the discharge,

$$\left(\frac{\dot{W}}{Q^3}\right)_2 = \left(\frac{\dot{W}}{Q^3}\right)_1$$

In other words, the power is proportional to the cubic power of the flow rate.

Solving for $P_2$ and substituting the given values yields

$$P_2 = P_1 \left(\frac{Q_2}{Q_1}\right)^3$$

$$= (10 \text{ kW}) \left(\frac{2 \frac{L}{s}}{1 \frac{L}{s}}\right)^3$$

$$= 80 \text{ kW}$$

**The answer is (C).**

**9.** The coefficient of velocity for sharp-edged orifices is 0.98. **[Orifices]**

From a water properties table, the density of water at 25°C is 997 kg/m³. **[Properties of Water (SI Metric Units)]**

Assuming a horizontal venturi meter, the formula can be simplified.

**Venturi Meters**

$$Q = \frac{C_v A_2}{\sqrt{1-(A_2/A_1)^2}} \sqrt{2g\left(\frac{P_1}{\gamma}+z_1-\frac{P_2}{\gamma}-z_2\right)}$$

$$= \frac{C_v A_2}{\sqrt{1-(A_2/A_1)^2}} \sqrt{2g\left(\frac{P_1}{\gamma}-\frac{P_2}{\gamma}\right)}$$

The specific weight of a liquid is

$$\gamma = \rho g$$

The area at the throat is

$$A_2 = \frac{\pi D_2^2}{4}$$

$$= \frac{\pi (3.5 \text{ cm})^2}{4}$$

$$= 9.6 \text{ cm}^2$$

The area of the cross section of the pipe is

$$A_1 = \frac{\pi D_1^2}{4}$$

$$= \frac{\pi (8 \text{ cm})^2}{4}$$

$$= 50 \text{ cm}^2$$

The density of mercury is $13.6 \times 10^3$ kg/m³, and the acceleration due to gravity is 9.81 m/s². The minimum reading on the manometer is 1 cm (0.01 m). The minimum pressure drop $(p_1 - p_2)$ can be computed from the formula

$$(p_1 - p_2) = \rho_{Hg} g \Delta h$$

Substituting and solving gives

$$Q = \frac{(0.98)(9.6 \text{ cm}^2)\left(\frac{1 \text{ m}}{100 \text{ cm}}\right)^2}{\sqrt{1-\left(\frac{9.62 \text{ cm}}{50 \text{ cm}}\right)^2}}$$

$$\times \sqrt{\frac{2}{997 \frac{\text{kg}}{\text{m}^3}} \left(13.6 \times 10^3 \frac{\text{kg}}{\text{m}^3}\right)\left(9.81 \frac{\text{m}}{\text{s}^2}\right)(0.01 \text{ m})}$$

$$= 1.6 \times 10^{-3} \text{ m}^3/\text{s}$$

**The answer is (A).**

**10.** The coefficient of the orifice meter, $C$, is given as 0.61. **[Orifices]**

The diameter of the orifice is given as 1 cm (0.01 m), and the density of water at 20°C is 998.2 kg/m³. **[Properties of Water (SI Metric Units)]**

The maximum flow rate, $Q$, can be found from the formula

Orifices

$$Q = CA_0\sqrt{2g\left(\frac{P_1}{\gamma} + z_1 - \frac{P_2}{\gamma} - z_2\right)}$$
$$= CA_0\sqrt{2g\left(\frac{P_1}{\gamma} - \frac{P_2}{\gamma}\right)}$$

The density of mercury is $13.6 \times 10^3$ kg/m³, and the maximum reading of the manometer is 10 cm (0.1 m). The pressure drop in the orifice meter is

$$(p_1 - p_2) = \rho_{Hg}g\Delta h$$

The specific weight of a liquid is

$$\gamma = \rho g$$

Substituting gives

$$Q = (0.61)\left(\frac{\pi(0.01 \text{ m})^2}{4}\right)$$
$$\times \sqrt{\frac{(2)\left(13.6 \times 10^3 \; \frac{\text{kg}}{\text{m}^3}\right)\left(9.81 \; \frac{\text{m}}{\text{s}^2}\right)(0.10 \text{ m})}{998.2 \; \frac{\text{kg}}{\text{m}^3}}}$$
$$= 2.5 \times 10^{-4} \text{ m}^3/\text{s}$$

**The answer is (D).**

**11.** The hydraulic radius is the cross-sectional area of flow divided by the wetted perimeter. Thus,

Manning's Equation
$$R_H = A/P$$

The cross-sectional area of flow is

$$A = (5 \text{ m})(2 \text{ m})$$
$$= 10 \text{ m}^2$$

The wetted perimeter is on three sides (i.e., all sides except the top).

$$P = 2 \text{ m} + 5 \text{ m} + 2 \text{ m}$$
$$= 9 \text{ m}$$

Therefore,

$$R_H = \frac{10 \text{ m}^2}{9 \text{ m}}$$
$$= 1.1 \text{ m}$$

**The answer is (A).**

**12.** The volumetric flow rate is

$$Q = \frac{10 \; \frac{\text{L}}{\text{s}}}{1000 \; \frac{\text{L}}{\text{m}^3}} = 0.01 \text{ m}^3/\text{s}$$

The velocity needed is

$$v = \frac{Q}{A} = \frac{0.01 \; \frac{\text{m}^3}{\text{s}}}{0.5 \text{ m}^2} = 0.02 \text{ m/s}$$

Use Manning's equation to find the slope needed to achieve this velocity.

$$v = (K/n)R_H^{2/3}S^{1/2}$$
$$S = \left(\frac{vn}{KR_H^{2/3}}\right)^2$$
$$= \left(\frac{\left(0.02 \; \frac{\text{m}}{\text{s}}\right)(0.15)}{(1.0)(0.15 \text{ m})^{2/3}}\right)^2$$
$$= 0.0001129 \quad (1.1 \times 10^{-4})$$

**The answer is (A).**

**13.** A hydraulic elements graph can be used to obtain the solution. [**Hydraulic-Elements Graph for Circular Sewers**]

The ratio of depth, $d$, to diameter, $D$, must be found, as follows.

$$\frac{d}{D} = \frac{20 \text{ cm}}{60 \text{ cm}} = 0.333$$

The velocity curve in the chart at that depth-to-diameter ratio reads 0.64, so that

$$\frac{v}{v_f} = 0.64$$

Solving for v and substituting the given value for $v_f$ yields

$$v = 0.64v_f$$
$$= (0.64)\left(1.1 \; \frac{\text{m}}{\text{s}}\right)$$
$$= 0.70 \text{ m/s}$$

**The answer is (A).**

**14.** A hydraulic elements graph can be used to obtain the solution. [**Hydraulic-Elements Graph for Circular Sewers**]

The ratio of depth, $d$, to diameter, $D$, must be found, as follows.

$$\frac{d}{D} = \frac{10 \text{ cm}}{30 \text{ cm}} = 0.333$$

The area curve in the chart at that depth-to-diameter ratio reads 0.28, so that

$$\frac{A}{A_f} = 0.28$$

The full-flow cross-sectional area is

$$A_f = \frac{\pi D^2}{4}$$

Solving for $A$ and substituting for $A_f$ yields

$$\begin{aligned} A &= 0.28 A_f \\ &= 0.28 \left(\frac{\pi D^2}{4}\right) \\ &= (0.28)\left(\frac{\pi (30 \text{ cm})^2}{4}\right) \\ &= 200 \text{ cm}^2 \quad (0.020 \text{ m}^2) \end{aligned}$$

**The answer is (D).**

**15.** The diameter of the pipe is 0.06 m, the velocity of flow is 10 m/s, the head loss is 0.5 m/m, and the density of water at 25°C is 997 kg/m³. [**Properties of Water (SI Metric Units)**]

The loss of power is given by

$$P = Q\rho h_f g$$

The volumetric flow rate is given by

$$Q = A\text{v} = \left(\frac{\pi D^2}{4}\right)\text{v}$$

Substitution gives

$$\begin{aligned} P &= Q\rho h_f g \\ &= \left(\frac{\pi}{4}(0.06 \text{ m})^2\right)\left(10 \frac{\text{m}}{\text{s}}\right)\left(997 \frac{\text{kg}}{\text{m}^3}\right)\left(0.5 \frac{\text{m}}{\text{m}}\right)\left(9.8 \frac{\text{m}}{\text{s}^2}\right) \\ &= 140 \text{ kg}\cdot\text{m}^2/\text{s}^3/\text{m} \quad (0.14 \text{ kW/m}) \end{aligned}$$

**The answer is (B).**

**16.** Since the thickness of the pipe is 0.5 cm, the water extends a distance of

$$\begin{aligned} h &= \text{height of water in tube above pipe} \\ &\quad + \text{pipe thickness} \\ &= 1 \text{ cm} + 0.5 \text{ cm} \\ &= 1.5 \text{ cm} \end{aligned}$$

The equation for the pitot tube is

$$\text{v}^2 = 2gh$$

Solving for v and substituting values gives

$$\text{v} = \sqrt{2gh} = \sqrt{(2)\left(980 \frac{\text{cm}}{\text{s}^2}\right)(1.5 \text{ cm})} = 54 \text{ cm/s}$$

**The answer is (C).**

**17.** The volumetric flow rate is

**Air Stripping**

$$\begin{aligned} Q_W \cdot C_{in} &= Q_A H' C_{in} \\ Q_W = Q_A H' &= \frac{Q_A H}{RT} \\ &= \frac{\left(1 \frac{\text{m}^3}{\text{s}}\right)\left(1000 \frac{\text{L}}{\text{m}^3}\right)\left(0.10 \frac{\text{atm}\cdot\text{L}}{\text{mol}}\right)}{\left(0.08206 \frac{\text{atm}\cdot\text{L}}{\text{mol}\cdot\text{K}}\right)(25°\text{C} + 273°)} \\ &= 4.09 \text{ L/s} \quad (4 \text{ L/s}) \end{aligned}$$

**The answer is (B).**

**18.** The primary treatment of wastewater is the removal of solids and particles.

**The answer is (D).**

**19.** Small particles generally follow Stokes' law.

**Settling Equations: Stokes' Law**

$$\text{v}_t = \frac{g(\rho_p - \rho_f)d^2}{18\mu}$$

Values for variables are $\rho_f = 1$ g/cc or 1000 kg/m³, $g = 9.81$ m/s², and $d = 5 \times 10^{-4}$ m.

$$\begin{aligned} \mu &= (0.87 \text{ cP})\left(0.001 \frac{\text{Pa}\cdot\text{s}}{\text{cP}}\right)\left(1 \frac{\frac{\text{N}}{\text{m}^2}}{\text{Pa}}\right)\left(1 \frac{\frac{\text{kg}\cdot\text{m}}{\text{s}^2}}{\text{N}}\right) \\ &= 8.7 \times 10^{-4} \text{ kg/m}\cdot\text{s} \end{aligned}$$

$$\rho_p = \left(1.8 \ \frac{\text{g}}{\text{cm}^3}\right)\left(\frac{1 \ \text{kg}}{1000 \ \text{g}}\right)\left(100 \ \frac{\text{cm}}{\text{m}}\right)^3$$
$$= 1800 \ \text{kg/m}^3$$

$$v_t = \frac{\left(9.81 \ \frac{\text{m}}{\text{s}^2}\right)\left(1800 \ \frac{\text{kg}}{\text{m}^3} - 1000 \ \frac{\text{kg}}{\text{m}^3}\right)(5\times 10^{-4} \ \text{m})^2}{(18)\left(8.7 \times 10^{-4} \ \frac{\text{kg}}{\text{m}\cdot\text{s}}\right)}$$
$$= 0.13 \ \text{m/s}$$

**The answer is (B).**

**20.** The 5-day BOD of the diluted wastewater is

$$y_5 = \text{DO}_{0d} - \text{DO}_{5d}$$
$$= 7.0 \ \frac{\text{mg}}{\text{L}} - 3.0 \ \frac{\text{mg}}{\text{L}}$$
$$= 4 \ \text{mg/L}$$

Since this wastewater is diluted by a factor of 10, the original wastewater has a 5-day BOD of

$$y_5 = (\text{dilution factor})(\text{BOD of diluted water})$$
$$= (10)\left(4 \ \frac{\text{mg}}{\text{L}}\right)$$
$$= 40 \ \text{mg/L}$$

**The answer is (D).**

**21.** Here, the kinetic constant is given and the ultimate BOD must be found.

**Microbial Kinetics: BOD Exertion**
$$\text{BOD}_t = L_o(1 - e^{-kt})$$

The BOD can also be found from the difference between the initial dissolved oxygen and the dissolved oxygen after 2 d. Therefore,

$$\text{BOD}_{2d} = \text{DO}_{0d} - \text{DO}_{2d}$$
$$= 8 \ \frac{\text{mg}}{\text{L}} - 6 \ \frac{\text{mg}}{\text{L}}$$
$$= 2 \ \text{mg/L}$$
$$= L_o\left(1 - e^{-\left(0.1 \frac{1}{\text{d}}\right)(2 \ \text{d})}\right)$$
$$= L_o(1 - e^{-0.2})$$

Solve for $L_o$.

$$L_o = \frac{2 \ \frac{\text{mg}}{\text{L}}}{1 - e^{-0.2}}$$
$$= 11 \ \text{mg/L}$$

**The answer is (C).**

**22.** Finer particles that cause turbidity must be treated with coagulants in order to form larger particles that can be removed by settling and filtration. High turbidity requires better filtration operation coupled with coagulation.

**The answer is (A).**

**23.** Total hardness is the concentration of calcium and magnesium ions expressed as an equivalent of calcium carbonate. The sodium cations do not contribute toward the hardness of the water. The atomic weight of calcium is 40 g/mol, the atomic weight of magnesium is 24 g/mol, and the molecular weight of calcium carbonate is 100 g/mol. [**Lime-Soda Softening Equations**]

Find the calcium carbonate equivalents of the calcium and magnesium ions.

$$20 \ \frac{\text{mg Ca}^{2+}}{\text{L}} = \left(\frac{20 \ \frac{\text{mg Ca}^{2+}}{\text{L}}}{40 \ \frac{\text{g Ca}^{2+}}{\text{mol}}}\right)\left(100 \ \frac{\text{g CaCO}_3}{\text{mol}}\right)$$
$$= 50 \ \text{mg/L CaCO}_3$$

$$20 \ \frac{\text{mg Mg}^{2+}}{\text{L}} = \left(\frac{20 \ \frac{\text{mg Mg}^{2+}}{\text{L}}}{24 \ \frac{\text{g Mg}^{2+}}{\text{mol}}}\right)\left(100 \ \frac{\text{g CaCO}_3}{\text{mol}}\right)$$
$$= 83 \ \text{mg/L CaCO}_3$$

Therefore, the total hardness is

$$\text{total hardness} = 50 \ \frac{\text{mg}}{\text{L CaCO}_3} + 83 \ \frac{\text{mg}}{\text{L CaCO}_3}$$
$$= 130 \ \text{mg/L CaCO}_3$$

**The answer is (C).**

**24.** Assume the density of water is 1000 kg/m³ and the viscosity of water is 0.87 cP, which is equivalent to $8.7 \times 10^{-4}$ kg/m·s. Calculate the Reynolds number.

**Filtration Equations: Head Loss Through Clean Bed**

$$\text{Re} = \frac{v_s \rho d}{\mu}$$

$$= \frac{\left(0.3 \, \frac{\text{m}}{\text{s}}\right)\left(1000 \, \frac{\text{kg}}{\text{m}^3}\right)(0.03 \, \text{m})}{8.7 \times 10^{-4} \, \frac{\text{kg}}{\text{m} \cdot \text{s}}}$$

$$= 10\,000$$

Calculate the friction factor.

$$f' = \frac{150(1-\eta)}{\text{Re}} + 1.75$$

$$= \frac{(150)(1-0.47)}{10\,000} + 1.75$$

$$= 1.8$$

Calculate the head loss.

$$h_f = \frac{f' L (1-\eta) v_s^2}{\eta^3 g d_p}$$

$$= \frac{(1.8)(20 \, \text{m})(1-0.47)\left(0.3 \, \frac{\text{m}}{\text{s}}\right)^2}{(0.47)^3 \left(9.81 \, \frac{\text{m}}{\text{s}^2}\right)(0.03 \, \text{m})}$$

$$= 56 \, \text{m}$$

**The answer is (C).**

**25.** The equilibrium constant for this reaction is

**Equilibrium Constant of a Chemical Reaction**

$$K_{eq} = \frac{[C]^c [D]^d}{[A]^a [B]^b}$$

$$= \frac{[H^+][CO_3^{-2}]}{[HCO_3^-]} = 5 \times 10^{-11} \, \text{mol/L}$$

The molar concentration of $CO_3^{-2}$ is the same as the molar concentration of $H^+$. Substitute and solve for $[HCO_3^-]$.

$$[HCO_3^-] = \frac{[H^+]^2}{5 \times 10^{-11} \, \frac{\text{mol}}{\text{L}}}$$

[H+] is related to pH, as follows.

**Acids, Bases, and pH**

$$\text{pH} = \log_{10}\left(\frac{1}{[H^+]}\right)$$

$$= 7.5$$

Solving for $[H^+]$ yields

$$[H^+] = 10^{-7.5} \, \text{mol/L}$$

Substitute to find $[HCO_3^-]$.

$$[HCO_3^-] = \frac{\left(10^{-7.5} \, \frac{\text{mol}}{\text{L}}\right)^2}{5 \times 10^{-11} \, \frac{\text{mol}}{\text{L}}}$$

$$= 2.0 \times 10^{-5} \, \text{mol/L}$$

**The answer is (C).**

**26.** The gas given out during the digestion process is methane.

**The answer is (B).**

**27.** First, balance the chemical equation.

$$C_8H_{18} + 12.5 O_2 \rightarrow 8 CO_2 + 9 H_2O$$

Find the molecular weight of octane.

$$\text{MW}_{\text{octane}} = (8)\left(12 \, \frac{\text{g}}{\text{mol}}\right) + (18)\left(1 \, \frac{\text{g}}{\text{mol}}\right)$$

$$= 114 \, \text{g/mol}$$

Next, find the molecular weight of oxygen.

$$\text{MW}_{O_2} = (2)\left(16 \, \frac{\text{g}}{\text{mol}}\right)$$

$$= 32 \, \text{g/mol}$$

$$\text{COD} = \left(12.5 \, \frac{\text{mol } O_2}{\text{mol octane}}\right) \left(\frac{32 \, \frac{\text{g } O_2}{\text{mol } O_2}}{114 \, \frac{\text{g octane}}{\text{mol octane}}}\right)$$

$$= 3.5 \, \text{g } O_2/\text{g octane}$$

**The answer is (A).**

**28.** Two molecules of oxygen are required for every molecule of ammonia. Find the molecular weight of oxygen and of ammonia.

$$\text{MW}_{O_2} = (2)\left(16 \frac{g}{mol}\right)$$
$$= 32 \text{ g/mol}$$

$$\text{MW}_{NH_3} = (1)\left(14 \frac{g}{mol}\right) + (3)\left(1 \frac{g}{mol}\right)$$
$$= 17 \text{ g/mol}$$

$$\text{mass}_{O_2} = 2(\text{mass}_{NH_3})\left(\frac{\text{MW}_{O_2}}{\text{MW}_{NH_3}}\right)$$

$$= (2)(100 \text{ kg NH}_3)\left(\frac{32 \frac{g\ O_2}{mol\ O_2}}{17 \frac{g\ NH_3}{mol\ NH_3}}\right)$$

$$= 380 \text{ kg O}_2$$

**The answer is (C).**

**29.** Reaction rate, $k$, and half-life are related.

Half-Life

$$k = \frac{0.693}{t_{1/2}}$$
$$= \frac{0.693}{21 \text{ d}}$$
$$= 0.033/\text{d}$$

**The answer is (A).**

**30.** Using the formula to solve for the reaction-rate constant at 5°C,

$$k(T) = (0.2 \text{ d}^{-1})(1.024^{5°C-20°C})$$
$$= 0.14/\text{d}$$

**The answer is (A).**

**31.** Using the operator $D$ for $d/dx$, the equation becomes

$$D^2 Y + 16 Y = 0$$

This can also be phrased as

$$(D^2 + 16) Y = 0$$

The roots of the characteristic equation are $r_1$ and $r_2$. [**Second-Order Linear Homogeneous Differential Equations with Constant Coefficients**]

The characteristic form of this equation is

$$(r^2 + r + 16) Ce^{rx} = 0$$

$Ce^{rx}$ cannot be zero, so

$$r + 16 = 0$$

This can also be expressed as

$$r = \frac{\pm\sqrt{-16x4}}{2} = \pm 4i$$

Solve the equation for $r$.

$$r = C_1 e^{(r_1 x)} + C_2 e^{(-r_2 x)}$$
$$= C_1 e^{(4ix)} + C_2 e^{(-4ix)}$$

$C_1$ and $C_2$ are constants.

$$r = C_1(\cos 4x + i\sin 4x) + C_2(\cos 4x - i\sin 4x)$$
$$= C_3 \cos 4x + C_4 \sin 4x$$

$\alpha$ and $\beta$ can replace $C_3$ and $C_4$.

$$\alpha\cos 4x + \beta\sin 4x$$

**The answer is (D).**

**32.** Use the Streeter-Phelps stream modeling equation, with $k_d = 0.1/\text{d}$, $k_a = 2.00/\text{d}$, and $S_0 = 10$ mg/L. [**Microbial Kinetics: Stream Modeling**]

$$t = \frac{\text{distance}}{v}$$
$$= \frac{160 \text{ km}}{\left(0.11 \frac{\text{km}}{\text{min}}\right)\left(60 \frac{\text{min}}{\text{h}}\right)\left(24 \frac{\text{h}}{\text{d}}\right)}$$
$$= 1.0 \text{ d}$$

The initial dissolved oxygen deficit in the mixing zone is

$$D_0 = DO_{\text{sat}} - DO_0$$
$$= 8.9 \frac{\text{mg}}{\text{L}} - 8.5 \frac{\text{mg}}{\text{L}}$$
$$= 0.4 \text{ mg/L}$$

The dissolved oxygen deficit after 160 km is

$$D = \left(\frac{k_d S_0}{k_a - k_d}\right)(e^{-k_d t} - e^{-k_a t}) + D_0 e^{-k_a t}$$

$$= \left(\frac{(0.1 \text{ d}^{-1})\left(10 \frac{\text{mg}}{\text{L}}\right)}{2 \text{ d}^{-1} - 0.1 \text{ d}^{-1}}\right)$$

$$\times \left(e^{(-0.1 \text{ d}^{-1})(1.0 \text{ d})} - e^{(-2 \text{ d}^{-1})(1.0 \text{ d})}\right)$$

$$+ \left(0.4 \frac{\text{mg}}{\text{L}}\right) e^{(-2 \text{ d}^{-1})(1.01 \text{ d})}$$

$$= 0.46 \text{ mg/L}$$

The dissolved oxygen concentration after 160 km is found using the equation

**Microbial Kinetics: Stream Modeling**

$$D = DO_{\text{sat}} - DO$$

Solving for $DO$,

$$DO = DO_{\text{sat}} - D$$

$$= 8.9 \frac{\text{mg}}{\text{L}} - 0.46 \frac{\text{mg}}{\text{L}}$$

$$= 8.4 \text{ mg/L}$$

**The answer is (D).**

**33.** Option A represents a zero-order reaction. [**Zero-Order Irreversible Reaction**]

For a zero-order reaction, concentration changes linearly with time. The concentration drop on the first day is

$$C - C_0 = 3.3 \frac{\text{g}}{\text{L}} - 10 \frac{\text{g}}{\text{L}} = -6.7 \text{ g/L}$$

If the reaction were zero order, it would drop by the same amount on the second day. It does not drop by that amount, so the reaction is not zero order, and A is not the solution.

Option B represents a first-order. [**First-Order Irreversible Reaction**] In a first-order reaction, $\ln(C/C_0)$ is a linear function of time. On the first day,

$$\ln \frac{3.3 \frac{\text{g}}{\text{L}}}{10 \frac{\text{g}}{\text{L}}} = -1.1$$

On the second day,

$$\ln \frac{2.0 \frac{\text{g}}{\text{L}}}{3.3 \frac{\text{g}}{\text{L}}} = -0.5$$

Therefore, the reaction is not first order.

Option C represents a second-order. [**Second-Order Irreversible Reaction**]

In a second-order reaction, $1/C$ is a linear function of time.

At $t = 0$ d,

$$\frac{1}{C} = 0.1 \text{ L/g}$$

At $t = 1$ d,

$$\frac{1}{C} = 0.3 \text{ L/g}$$

At $t = 2$ d,

$$\frac{1}{C} = 0.5 \text{ L/g}$$

Each day, the value for $1/C$ increases by 0.2 L/g, so the reaction is second order, with a rate constant of 0.2 L/g·d.

The rate expression for a second-order reaction is

$$\frac{dC}{dt} = -kC^2$$

Integrating gives

$$\int_{C_0}^{C} \frac{dC}{C^2} = -k \int_{0 \text{ d}}^{t} dt$$

$$-\frac{1}{C} + \frac{1}{C_0} = -kt$$

Solving for $C$ gives

$$C = \left(kt + \frac{1}{C_0}\right)^{-1}$$

$$= \left(\left(0.2 \frac{\text{L}}{\text{g·d}}\right)t + \frac{1}{10 \frac{\text{g}}{\text{L}}}\right)^{-1}$$

**The answer is (C).**

**34.** At $t=0$ d, the concentration is equal to $C_0$. At $t=13$ d, the concentration is $0.16 C_0$. In a first-order reaction,

**First-Order Irreversible Reaction**
$$-dC_A/dt = kC_A$$
$$\frac{dC_A}{dt} = -kC_A$$

Integrating and solving for $k$ gives

$$\int_{C_0}^{0.16 C_0} \frac{dC_A}{C_A} = -k \int_{0\,\text{d}}^{13\,\text{d}} dt$$

$$\ln \frac{0.16 C_0}{C_0} = -k(13\,\text{d})$$

$$k = -\frac{\ln 0.16}{13\,\text{d}}$$

$$= 0.14/\text{d}$$

**The answer is (D).**

**35.** From a table for air-cloth ratio for baghouses, the value for limestone in a woven fabric shaker baghouse is $0.8\ \text{m}^3/\text{min·m}^2$. [**Baghouse**]

$$\frac{50\ \dfrac{\text{m}^3}{\text{min}}}{0.8\ \dfrac{\text{m}^3}{\text{min·m}^2}} = 60\ \text{m}^2$$

**The answer is (D).**

**36.** A Gaussian model of atmospheric dispersion is used for this problem. [**Atmospheric Dispersion Modeling (Gaussian)**]

The concentration at ground level directly under the centerline of a plume from a stack can be found by setting the horizontal distance from the plume centerline, $y$, and the vertical distance above ground level, $z$, equal to zero.

Substituting zero for $z$ and $y$ in the atmospheric dispersion modeling equation gives

$$C = \left(\frac{Q}{2\pi u \sigma_y \sigma_z}\right) e^{\frac{0^2}{2\sigma_y^2}} \left( e^{-\frac{(0-H)^2}{2\sigma_z^2}} + e^{-\frac{(0+H)^2}{2\sigma_z^2}} \right)$$

$$= \left(\frac{Q}{2\pi u \sigma_y \sigma_z}\right) \left( e^{-\frac{H^2}{2\sigma_z^2}} + e^{-\frac{H^2}{2\sigma_z^2}} \right)$$

$$= \left(\frac{Q}{\pi u \sigma_y \sigma_z}\right) e^{-\frac{H^2}{2\sigma_z^2}}$$

$$= \left(\frac{580\ \dfrac{\mu\text{g}}{\text{s}}}{\pi \left(2\ \dfrac{\text{m}}{\text{s}}\right)(40\ \text{m})(24\ \text{m})}\right) e^{-\frac{(80\ \text{m})^2}{(2)(24\ \text{m})^2}}$$

$$= 3.7 \times 10^{-4}\ \mu\text{g}/\text{m}^2$$

**The answer is (C).**

**37.** First find $X_0$ for oxygen, using the mass fraction of oxygen in air, $y_{O_2}$, and the total concentration of air at sea level.

$$X_{0,O_2} = y_{O_2} X_{0,\text{total}}$$
$$= (0.21)\left(1.2\ \frac{\text{kg}}{\text{m}^3}\right)\left(1000\ \frac{\text{g}}{\text{kg}}\right)$$
$$= 250\ \text{g}/\text{m}^3$$

Substitute this value into the relationship for finding $X_z$.

$$X_{2\,\text{km}} = \left(250\ \frac{\text{g}}{\text{m}^3}\right) e^{-\frac{2\,\text{km}}{8.4\,\text{km}}}$$
$$= 200\ \text{g}/\text{m}^3$$

**The answer is (B).**

**38.** This is a mass balance problem. The molecular weight of carbon dioxide is

$$\text{MW}_{CO_2} = 12\ \frac{\text{g}}{\text{mol}} + (2)\left(16\ \frac{\text{g}}{\text{mol}}\right)$$
$$= 44\ \text{g}/\text{mol}$$

If $y_{CO_2}$ is the mass fraction of $CO_2$ in the air, the total mass of carbon dioxide in the atmosphere is

$$m_{total, CO_2} = y_{CO_2} m_{total, air}$$

$$= \left(\frac{360 \text{ CO}_2 \text{ molecules}}{10^6 \text{ air molecules}}\right)\left(44 \frac{\text{g CO}_2}{\text{mol CO}_2}\right)$$

$$\times \left(\frac{1 \text{ mol air}}{29 \text{ g air}}\right)(5.2 \times 10^{18} \text{ kg air})$$

$$= 2.8 \times 10^{15} \text{ kg CO}_2$$

To find the flux,

$$F_{in} = F_{out} = \frac{M}{\tau}$$

$$= \left(\frac{2.8 \times 10^{15} \text{ kg CO}_2}{100 \text{ yr}}\right)\left(\frac{1 \text{ yr}}{365 \text{ d}}\right)$$

$$= 7.7 \times 10^{10} \text{ kg/d}$$

**The answer is (C).**

**39.** A Gaussian model of atmospheric dispersion is used for this problem. [**Atmospheric Dispersion Modeling (Gaussian)**]

First find the vertical distance between the top of the stack and the centerline of the plume where $x$, the downwind distance from the stack, is 0.4 km.

$$\Delta h = \frac{(1.6)\left(55 \frac{\text{m}^4}{\text{s}^3}\right)^{1/3} (400 \text{ m})^{2/3}}{3 \frac{\text{m}}{\text{s}}}$$

$$= 110 \text{ m}$$

The height of the plume's centerline above the ground is the stack height plus the height of the plume above the top of the stack.

$$h + \Delta h = 20 \text{ m} + 110 \text{ m}$$
$$= 130 \text{ m}$$

**The answer is (B).**

**40.** Use the formula for the stripper packing height.

**Air Stripping: Stripper Packing Height = Z**
$$Z = HTU \times NTU$$

HTU is given as 3 m, $H'$ is 10, $Q_A$ is $10^{-1}$ m³/s, $Q_W$ is 0.01 m³/s, $C_{in}$ is 100 ppm, and $C_{out}$ is 0.5 ppm.

$$NTU = \frac{R}{R-1} \ln \frac{\left(\frac{C_{in}}{C_{out}}\right)(R-1)+1}{R}$$

Here,

$$R = \frac{H'Q_A}{Q_W}$$

$$= \frac{(10)\left(10^{-1} \frac{\text{m}^3}{\text{s}}\right)}{10^{-2} \frac{\text{m}^3}{\text{s}}}$$

$$= 100$$

Plugging in the appropriate values gives

$$NTU = \frac{100}{100-1} \ln \frac{\left(\frac{100 \text{ ppm}}{0.5 \text{ ppm}}\right)(100-1)+1}{100}$$

$$= 5.3$$

$$Z = NTU \times HTU$$
$$= 5.3 \times 3 \text{ m}$$
$$= 16 \text{ m}$$

**The answer is (B).**

**41.** Increasing the cyclone's efficiency corresponds to decreasing the diameter of the particles that are collected with 50% efficiency ($d_{pc}$). The equation for calculating cyclone collection efficiency is

**Cyclone**

$$\eta = \frac{1}{1 + (d_{pc}/d_p)^2}$$

If a particle of diameter 20 μm is to have a collection efficiency of 80%, then $d_{pc}$ for the cyclone must be

$$d_{pc} = d_p \sqrt{\frac{1}{\eta} - 1}$$

$$= 20 \text{ μm} \sqrt{\frac{1}{0.8} - 1}$$

$$= 10 \text{ μm}$$

Alternatively, this value could be found using a cyclone collection efficiency chart. [**Cyclone Collection Efficiency**]

An equation for determining the diameter of particles collected at 50% efficiency is

$$d_{pc} = \sqrt{\frac{9\mu W}{2\pi N_e v_i (\rho_p - \rho_g)}}$$

In this equation, $d_{pc}$ is given in meters, $\mu$ is the viscosity of the gas in kg/m/s, $W$ is the inlet width of the cyclone in meters, $N_e$ is the number of effective turns the gas makes in the cyclone, $v_i$ is the inlet velocity into the cyclone in m/s, $\rho_p$ is the density of the particles in kg/m³, and $\rho_g$ is the density of the gas in kg/m³. Therefore, $d_{pc}$ is proportional to the square root of the inverse of the cyclone's inlet velocity.

An inlet velocity of 6 m/s corresponds to a $d_{pc}$ of 20 μm, and the new velocity must correspond to a $d_{pc}$ of 10 μm, so

$$\frac{10\ \mu\mathrm{m}}{20\ \mu\mathrm{m}} = \frac{\sqrt{\dfrac{1}{v_i}}}{\sqrt{\dfrac{1}{6\ \dfrac{\mathrm{m}}{\mathrm{s}}}}}$$

$$\left(\frac{1}{6\ \dfrac{\mathrm{m}}{\mathrm{s}}}\right)\left(\frac{1}{2}\right)^2 = \frac{1}{v_i}$$

$$v_i = 24\ \mathrm{m/s}$$

### The answer is (B).

**42.** Lowering the leak rate definition increases the annualized cost of the leak detection and repair program by $86,000/yr.

The value of the product not lost to fugitive emissions as a result of the program is

value of product not lost =
(product value) × (emission reduction amount)

$$= \left(4.20\ \frac{\$}{\mathrm{L}}\right)\left(195\ \frac{\mathrm{Mg}}{\mathrm{yr}}\right)\left(\frac{1\ \mathrm{L}}{14\ \mathrm{kg}}\right)\left(1000\ \frac{\mathrm{kg}}{\mathrm{Mg}}\right)$$

$$= \$58{,}500/\mathrm{yr}$$

Therefore, the net annualized cost of lowering the leak definition is

net annualized cost = (increase to annualized cost)
×(value of product not lost)

$$= 86{,}000\ \frac{\$}{\mathrm{yr}} - 58{,}500\ \frac{\$}{\mathrm{yr}}$$

$$= \$27{,}500/\mathrm{yr}$$

### The answer is (A).

**43.** Determine the atmospheric stability class for the given conditions. Atmospheric stability on a summer day with a few broken clouds and a wind speed of 4 m/s is moderately to slightly unstable (B − C). [**Atmospheric Stability Under Various Conditions**]

Thus, the average of the values at stabilities B and C is

$$C_{\mathrm{ave}} = \frac{C_{\mathrm{B}} + C_{\mathrm{C}}}{2}$$

$$= \frac{2.0 \times 10^{-4}\ \dfrac{\mathrm{g}}{\mathrm{m}^3} + 1.6 \times 10^{-4}\ \dfrac{\mathrm{g}}{\mathrm{m}^3}}{2}$$

$$= 1.8 \times 10^{-4}\ \mathrm{g/m}^3$$

### The answer is (D).

**44.** For gases, ppb is on a volume basis. For gases at atmospheric pressures and temperatures, the ideal gas law applies, and this is equivalent to a molar basis. The chemical equation shows a 1:1 molar ratio of ozone to carbon monoxide, so the effect is to increase the concentration of ozone by

$$(50\ \mathrm{ppb\ CO})\left(\frac{1\ \mathrm{mol\ O}_3}{1\ \mathrm{mol\ CO}}\right) = 50\ \mathrm{ppb\ O}_3$$

### The answer is (C).

**45.** The leachability of specific compounds at greater than threshold concentrations determines whether a waste is characteristically RCRA hazardous for toxicity.

### The answer is (C).

**46.** The strict, joint, and several provisions of the Superfund legislation mean that a company might be held responsible for all of the cleanup costs associated with a site, even if the company's activities were legal and contributed only a small fraction of the contamination at the site.

### The answer is (D).

**47.** The components shown in the table are typical of municipal solid waste. The total mass of all components in the table is

mass % components other than water
$= 3.2\% + 0.2\% + 1.2\% + 18.9\% + 3.8\%$
$\quad + 15.8\% + 33.0\%$
$= 76.1\%$

Therefore, the mass percent of water is

mass % water $= 100\% - 76.1\%$
$= 23.9\%$

### The answer is (B).

**48.** The formula for destruction and removal efficiency during incineration, $DRE$, is

*Incineration*

$$DRE = \frac{W_{\mathrm{in}} - W_{\mathrm{out}}}{W_{\mathrm{in}}} \times 100\%$$

Solve for $W_{out}$.

$$W_{out} = W_{in} - (DRE)W_{in}$$
$$= 1.1 \times 10^{-3} \frac{kg}{s} - (0.9999)\left(1.1 \times 10^{-3} \frac{kg}{s}\right)$$
$$= 1.1 \times 10^{-7} \text{ kg/s}$$

**The answer is (A).**

**49.** Incinerability is determined from

$$I = C + \frac{100 \frac{kcal}{g}}{H}$$

For acetonitrile,

$$I = 3.2 + \frac{100 \frac{kcal}{g}}{7.37 \frac{kcal}{g}} = 17$$

For benzene,

$$I = 5.7 + \frac{100 \frac{kcal}{g}}{10.03 \frac{kcal}{g}} = 16$$

For naphthalene,

$$I = 4.7 + \frac{100 \frac{kcal}{g}}{9.62 \frac{kcal}{g}} = 15$$

For vinyl chloride,

$$I = 3.0 + \frac{100 \frac{kcal}{g}}{4.45 \frac{kcal}{g}} = 25$$

In order of difficulty to incinerate (increasing incinerability index), the compounds are naphthalene, benzene, acetonitrile, and vinyl chloride.

**The answer is (D).**

**50.** An artesian aquifer is one where the water pressure at the top of the water in the aquifer is greater than the atmospheric pressure. Therefore, the water level of a well drilled into an artesian aquifer will be above the level of the top of the water in the aquifer.

**The answer is (B).**

**51.** For a first-order reaction,

**First-Order Irreversible Reaction**

$$-dC_A/dt = kC_A$$
$$\frac{dC_A}{dt} = -kC_A$$

Integrate to get

$$\int_{C_{A,0}}^{C_A} \frac{1}{C_A} dC_A = -k \int_0^t dt$$
$$\ln \frac{C_A}{C_{A,0}} = -kt$$

Solve for $k$.

$$k = \frac{1}{t} \ln \frac{C_{A,0}}{C_A}$$

Solve the Arrhenius equation for $T$.

**Chemical Reaction Engineering: Nomenclature**

$$k = Ae^{-E_a/\overline{R}T}$$
$$\ln \frac{k}{A} = -\frac{E_a}{\overline{R}T}$$
$$T = -\frac{E_a}{\overline{R} \ln \frac{k}{A}}$$

Substitute for $k$.

$$T = -\frac{E_a}{\overline{R} \ln \frac{\frac{1}{t} \ln \frac{C_{A,0}}{C_A}}{A}}$$

$$= \frac{-49 \times 10^3 \frac{cal}{mol}}{\left(1.987 \frac{cal}{mol \cdot K}\right) \ln \frac{\frac{1}{1.2} \ln \frac{1}{1 \times 10^{-4}}}{2.90 \times 10^{12} \frac{1}{s}}}$$

$$= 930 K$$

**The answer is (B).**

**52.** Option B indicates a linear graph. An integration of the material balance equation using Darcy's law gives a formula that can be interpreted as a conical surface for flow of water.

### The answer is (C).

**53.** Use Darcy's equation. First, find the hydraulic gradient.

$$\frac{dh}{dx} = -0.03$$

Darcy's equation is

**Darcy's Law**
$$Q = -KA(dh/dx)$$

Solve for $K$.

$$K = \frac{Q}{\frac{dh}{dx} A}$$

$$= \frac{\left(0.1 \, \frac{\text{L}}{\text{s}}\right)\left(\frac{1 \, \text{m}^3}{1000 \, \text{L}}\right)}{(0.03)(11 \, \text{m}^2)}$$

$$= 3 \times 10^{-4} \, \text{m/s}$$

### The answer is (A).

**54.** The precision of an instrument reflects the number of significant digits in an instrument reading. The accuracy of an instrument reflects how close the instrument reading is to the true value of what is being measured.

### The answer is (C).

**55.** Using $D$ to represent $d/dx$, the differential equation can be written as

$$(D^2 + 1)Y = 4$$

$(D^2 + 1)Y = 0$ represents the general differential equation form.

$$Y = Ae^{ix} + \beta e^{-ix} = C\cos x + D\sin x$$

$C$ and $D$ are general constants, so this can be rewritten as

$$Y = A\cos + \beta\sin + \text{other terms that are functions of } X.$$

### The answer is (D).

**56.** Stokes' law is

**Settling Equations: Stokes' Law**
$$v_t = \frac{g(\rho_p - \rho_f)d^2}{18\mu}$$

The gravitational constant, the difference between the particle density and the air density, and the viscosity of air are the same for both particle sizes. Therefore, the rates of deposition can be compared as follows.

$$\frac{v_{t,a}}{v_{t,b}} = \frac{(d_a)^2}{(d_b)^2}$$

$$\frac{v_{t,15 \, \mu\text{m particle}}}{v_{t,10 \, \mu\text{m particle}}} = \frac{(15 \, \mu\text{m})^2}{(10 \, \mu\text{m})^2} = 2.3$$

The deposition rate of the larger particle is 2.3 times the rate of the smaller particle.

### The answer is (C).

**57.** The concentration of ammonia is $8.1 \times 10^{-3}$ moles per liter of solution. The solution is essentially 100% water. The molecular weight of water ($H_2O$) is

$$\text{MW}_{H_2O} = 2(\text{MW}_H) + \text{MW}_O$$

$$= (2)\left(1.0079 \, \frac{\text{g}}{\text{mol}}\right) + 15.999 \, \frac{\text{g}}{\text{mol}}$$

$$= 18.0148 \, \text{g/mol}$$

The number of moles of water in a liter of water is

$$n_W = \frac{m}{\text{MW}_{H_2O}} = \frac{\rho V}{\text{MW}_{H_2O}}$$

$$= \frac{\left(1000 \, \frac{\text{kg}}{\text{m}^3}\right)(1 \, \text{L})\left(1000 \, \frac{\text{g}}{\text{kg}}\right)}{\left(18.0148 \, \frac{\text{g}}{\text{mol}}\right)\left(1000 \, \frac{\text{L}}{\text{m}^3}\right)}$$

$$= 55.51 \, \text{mol}$$

The mole fraction of ammonia in the solution is

$$x = \frac{n_{\text{ammonia}}}{n_W} = \frac{8.1 \times 10^{-3} \, \text{mol}}{55.51 \, \text{mol}} = 1.459 \times 10^{-4}$$

Use Henry's law.

**Henry's Law at Constant Temperature**
$$P_i = hx_i$$
$$= (0.62 \, \text{atm})(1.459 \times 10^{-4})$$
$$= 9.047 \times 10^{-5} \, \text{atm} \quad (9.0 \times 10^{-5} \, \text{atm})$$

### The answer is (A).

**58.** The savings in landfill costs due to the use of mechanical scrapers is the volume of cling losses reduced by the use of mechanical scrapers, doubled because each liter of conditioner rinsed to the wastewater treatment unit results in 2 L of sludge, multiplied by the cost of trucking the sludge to a landfill.

$$\text{landfill cost savings} = (2)(\text{cling loss}_{\text{before}} - \text{cling loss}_{\text{after}})$$
$$\times (\text{landfill cost})$$
$$= (2)\left(2.6 \frac{\text{L}}{\text{batch}} - 0.8 \frac{\text{L}}{\text{batch}}\right)$$
$$\times \left(3 \frac{\text{batch}}{\text{d}}\right)\left(365 \frac{\text{d}}{\text{yr}}\right)$$
$$\times \left(1.50 \frac{\$}{\text{L}}\right)$$
$$= \$5900/\text{yr}$$

**The answer is (D).**

**59.** The equation for the volumetric flow rate is

$$Q = \frac{C_V A_2}{\sqrt{1 - \left(\frac{A_2}{A_1}\right)^2}} \sqrt{\frac{2(P_1 - P_2)}{\rho}}$$

The ratio of areas can be simplified to

$$\frac{A_2}{A_1} = \frac{\frac{\pi D_2^2}{4}}{\frac{\pi D_1^2}{4}} = \left(\frac{D_2}{D_1}\right)^2 = \left(\frac{2 \text{ in}}{5 \text{ in}}\right)^2$$
$$= 0.16$$

Convert the density to kilograms per cubic meter.

$$\rho = \left(1 \frac{\text{g}}{\text{cm}^3}\right)\left(10^3 \frac{\text{kg}}{\text{mg}}\right) = 10^3 \text{ kg/m}^3$$

Find the area of the throat in meters.

$$A_2 = \frac{\pi D_2^2}{4} = \left(\frac{\pi (2 \text{ in})^2}{4}\right)\left(0.0254 \frac{\text{m}}{\text{in}}\right)^2 = 0.00203 \text{ m}^2$$

Convert the pressure drop to newtons per square meter.

$$(P_1 - P_2) = \frac{6 \text{ mm Hg}}{\left(760 \frac{\text{mm Hg}}{\text{atm}}\right)\frac{\left(1.013 \times 10^5 \frac{\text{N}}{\text{m}^2}\right)}{1 \text{ atm}}}$$
$$= 799.7 \text{ N/m}^2$$

For a rounded venturi meter, the roughness coefficient is 0.98. The volumetric flow rate is

$$Q = \frac{(0.98)(0.00203 \text{ m}^2)}{\sqrt{1 - 0.16}} \sqrt{\frac{(2)\left(799.7 \frac{\text{N}}{\text{m}^2}\right)}{10^3 \frac{\text{kg}}{\text{m}^3}}}$$
$$= 0.0027 \text{ m}^3/\text{s}$$

**The answer is (C).**

**60.** The drinking water unit risk multiplied by the concentration in water multiplied by the number of exposed people is the number of predicted excess cancer cases.

$$\text{excess cancer cases} = C(\text{unit risk})(\text{exposed population})$$
$$= (2 \text{ ppb})\left(\frac{1 \frac{\mu g}{L}}{1 \text{ ppb}}\right)$$
$$\times \left(5 \times 10^{-5} \left(\frac{\mu g}{L}\right)^{-1}\right)$$
$$\times (10^6 \text{ people})$$
$$= 100 \text{ cases}$$

**The answer is (B).**

**61.** The fraction of the dose is determined from

$$\text{fraction} = \frac{\text{dose}_{10 \text{ m}}}{\text{dose}_{1 \text{ m}}} = \frac{(1 \text{ m})^2}{(10 \text{ m})^2} = 0.01$$

**The answer is (C).**

**62.** The formula for radiation half-life is

**Half-Life**

$$N = N_0 e^{-0.693 t/\tau}$$

The fraction remaining is $N/N_0$. Substituting values for half-life and time yields

$$\frac{N}{N_0} = e^{-\frac{(0.693)(6 \text{ hr})(60 \text{ min/hr})}{68.3 \text{ min}}}$$
$$= 0.026 \quad (2.6\%)$$

**The answer is (B).**

**63.** The work done on the engine is

**Closed Thermodynamic System**

$$w_b = \int P \, dv$$
$$= \int_{0.8 \text{ m}^3}^{0.6 \text{ m}^3} 1.1 V \, dv$$
$$= (1.1)\left(\frac{V^2}{2}\right)\Bigg|_{0.6 \text{ m}^3}^{0.8 \text{ m}^3}$$
$$= 0.154 \text{ atm·m}^3$$

**The answer is (B).**

**64.** The equation for the product of the partial pressure and the volume of carbon monoxide (CO) in the room is

$$PV = nRT$$

$n$ is the number of moles of CO in the room, which can be found from the equation shown.

$$n = \frac{m}{M}$$

$m$ is the mass of the CO, and $M$ is the molecular weight of the CO. The equation for the product of the partial pressure and the volume of CO can be rearranged to solve for the mass as shown.

$$PV = nRT = \frac{m}{M}RT$$
$$m = \frac{PVM}{RT}$$

Convert the partial pressure of the CO to units of atmospheres.

$$P = (1 \text{ mm Hg})\left(\frac{1 \text{ atm}}{760 \text{ mm Hg}}\right) = 0.0013 \text{ atm}$$

The absolute temperature in the room is

**Temperature Conversions**
$$T = 29°C + 273.15° = 302.15 \text{K}$$

The molecular weight of CO is

$$M = (1)\left(16 \, \frac{\text{kg}}{\text{kmol}}\right) + (1)\left(12 \, \frac{\text{kg}}{\text{kmol}}\right) = 28 \text{ kg/kmol}$$

The mass of CO in the room is

$$m = \frac{PVM}{RT}$$
$$= \frac{(0.0013 \text{ atm})(2000 \text{ L})\left(28 \, \frac{\text{g}}{\text{g·mol}}\right)}{\left(0.08206 \, \frac{\text{L·atm}}{\text{mol·K}}\right)(302.15\text{K})}$$
$$= 2.97 \text{ g} \quad (3.0 \text{ g})$$

**The answer is (B).**

**65.** Find the future worth of the initial expense. Interpolating from interest rate tables, the value of $F/P$ for a time of 15 years and an interest rate of 5% is the average of the values for an interest rate of 4% and an interest rate of 6%. **[Interest Rate Tables]**

$$(\$15,000,000)\left(\frac{F}{P}\right)\Bigg|_{15 \text{ yr},5\%} = \frac{(\$15,000,000)}{2} \times (1.8009 + 2.3966)$$
$$= \$31,480,000$$

The net annual income each year is $3,000,000 − $1,000,000 = $2,000,000.

Find the future worth of the net annual income. Interpolating from interest rate tables, the value of $F/A$ for a time of 15 years and an interest rate of 5% is the average of the values for an interest rate of 4% and an interest rate of 6%. The future worth of the net annual income is

$$(\$2,000,000)\left(\frac{F}{A}\right)\Bigg|_{15 \text{ yr},5\%} = \frac{(\$2,000,000)}{2} \times (20.0236 + 23.2766)$$
$$= \$43,290,000$$

The net future worth is

$$\$15,000,000 + \$43,290,000$$
$$-\$31,480,000 = \$26,810,000 \quad (\$27,000,000)$$

**The answer is (C).**

**66.** The benefits are $15,000,000 each year for 20 years. Find the present worth of the benefits. The present worth is equal to is $(\$15,000,000)(P/A)$. From interest rate tables, the value of $P/A$ for a time of 20 years and a 5% interest rate is 12.533. **[Interest Rate Tables]**

The present worth of the benefits is

$$(\$15,000,000)(12.533) = \$187,933,000$$

Find the present worth of the uniform expenses. The present worth is equal to ($1,000,000)($P/A$), and as with the calculation for the present worth of the benefits, the value of $P/A$ is 12.533. The present worth of the uniform expenses is

$$(\$1,000,000)(12.533) = \$12,533,000$$

The initial investment is $100,000,000, so the total present worth of costs is

$$\$100,000,000 + \$12,533,000 = \$112,533,000$$

The benefit-cost ratio is

$$\frac{\$187,933,000}{\$112,533,000} = 1.67 \quad (1.7)$$

**The answer is (B).**

**67.** At low pressure like 1 atm, the volume percentage is identical to the molar percentage. The mass of each component of the mixture is equal to the product of the number of moles of the component and the component's molecular weight, and the mass percentage is equal to the mass of the component divided by the total mass of the mixture. Using 100 mol of the gas mixture as a basis for calculation, there are 60 mol of nitrogen ($N_2$), 30 mol of oxygen ($O_2$), 5 mol of carbon monoxide (CO), and 5 mol of carbon dioxide ($CO_2$). From the atomic masses given in the periodic table of elements, the molecular weight of $N_2$ is 28 g/mol, the molecular weight of $O_2$ is 32 g/mol, the molecular weight of CO is 28 g/mol, and the molecular weight of $CO_2$ is 44 g/mol. The mass of $N_2$ in the mixture is

$$m_{N_2} = n_{N_2}MW_{N_2} = (60 \text{ mol})\left(28 \ \frac{\text{g}}{\text{mol}}\right) = 1680 \text{ g}$$

The mass of $O_2$ in the mixture is

$$m_{O_2} = n_{O_2}MW_{O_2} = (30 \text{ mol})\left(32 \ \frac{\text{g}}{\text{mol}}\right) = 960 \text{ g}$$

The mass of CO in the mixture is

$$m_{CO} = n_{CO}MW_{CO} = (5 \text{ mol})\left(28 \ \frac{\text{g}}{\text{mol}}\right) = 140 \text{ g}$$

The mass of $CO_2$ in the mixture is

$$m_{CO_2} = n_{CO_2}MW_{CO_2} = (5 \text{ mol})\left(44 \ \frac{\text{g}}{\text{mol}}\right) = 220 \text{ g}$$

The total mass of the mixture is 1680 g + 960 g + 140 g + 220 g = 3000 g. The mass percentage of $CO_2$ in the mixture is

$$\left(\frac{220 \text{ g}}{3000 \text{ g}}\right) \times 100\% = 7.33\% \quad (7.3\%)$$

**The answer is (C).**

**68.** Magnesium hydroxide has a chemical formula of $Mg(OH)_2$ and is a precipitating solid, as can be seen from lime-soda softening equations. [**Lime-Soda Softening Equations**]

The equation for the solubility product constant for slightly soluble substances is

**Chemistry: Definitions**

$$K_{SP} = [A^+]^m[B^-]^n$$

The related chemical equation is

$$A_mB_n \rightleftharpoons mA^{n+} + nB^{m-}$$

In this case, $A$ is Mg, $m$ is 1, $B$ is OH, and $n$ is 2.

$$Mg(OH)_2 \rightleftharpoons Mg^{2+} + 2OH^-$$

$$K_{SP} = [Mg^{2+}][OH^-]^2$$

Stoichiometry gives a relationship between the concentration of $Mg^{2+}$ and $OH^-$ at saturation.

$$[OH^-] = 2[Mg^{2+}]$$

The equation for pOH where concentration is in units of mol/L is

$$pOH = -\log[OH^-]$$

Also, pH and pOH are related as shown.

$$pOH + pH = 14$$

Substituting values and solving for the concentration of magnesium ions yields

$$K_{SP} = [Mg^{2+}](2[Mg^{2+}])^2 = 4[Mg^{2+}]^3$$

$$[Mg^{2+}] = \left(\frac{K_{SP}}{4}\right)^{1/3}$$

The pH is

$$\begin{aligned}
\text{pH} &= 14 - \text{pOH} = 14 + \log[\text{OH}^-] \\
&= 14 + \log(2[\text{Mg}^{2+}]) \\
&= 14 + \log\left(2\left(\frac{K_{\text{SP}}}{4}\right)^{1/3}\right) \\
&= 14 + \log\left((2)\left(\frac{2 \times 10^{-11} \frac{\text{mol}^3}{\text{L}^3}}{4}\right)^{1/3}\right) \\
&= 10.5
\end{aligned}$$

**The answer is (D).**

**69.** The degree of a carbon atom is the number of other carbon atoms it is bonded to. The carbon labeled 4 is only bonded to one other carbon atom, so it is degree 1.

**The answer is (A).**

**70.** The hydrogen attached to the carbon atom labeled 1 has the weakest carbon-to-hydrogen bond, because this carbon atom can stabilize itself with the three other carbon atoms it is attached to if its carbon-to-hydrogen bond is broken. Of the remaining labeled carbon atoms, the carbon atom labeled 4 can share the strain of losing one of its carbon-to-hydrogen bonds with only one other carbon atom, and the carbon atom labeled 2 can only share the strain of losing one of its carbon to hydrogen bonds with two other carbon atoms. There is no hydrogen attached to the carbon atom labeled 3, so halogenation cannot occur at this site.

**The answer is (A).**

**71.** The first step in International Union of Pure and Applied Chemistry nomenclature is to identify the longest carbon chain. In the case of this molecule, the longest chain is five carbons long, so it is a pentane. There are three methyl groups attached to this carbon chain, so it is a trimethylpentane, which means either option B or C could be correct. The carbons of the parent chain have to be numbered so that the substituents have the lowest possible numbers. In order to achieve this, the carbon with two methyls attached is numbered two, and the carbon with one methyl attached is numbered four. The IUPAC name of this molecule is 2,2,4-trimethylpentane.

**The answer is (C).**

**72.** The density of water at various temperatures is given in tables of properties of water. The relationship between temperature and density is not exactly linear, but interpolation to find the density of water at 6°C and at 45°C is adequate. [**Properties of Water (SI Metric Units)**]

$$\rho_{6°C} = \rho_{5°C} + \frac{(6°C - 5°C)(\rho_{10°C} - \rho_{5°C})}{10°C - 5°C}$$

$$\rho_{45°C} = \rho_{40°C} + \frac{(45°C - 40°C)(\rho_{50°C} - \rho_{40°C})}{50°C - 40°C}$$

Mass is density multiplied by volume. The mass of 250 mL of water at 6°C minus the mass of 250 ml of water at 45°C is

$$\begin{aligned}
\Delta m &= 250 \text{ mL}(\rho_{6°C} - \rho_{45°C}) \\
&= 250 \text{ mL}\left(\begin{array}{c} \rho_{5°C} + \frac{(6°C - 5°C)(\rho_{10°C} - \rho_{5°C})}{10°C - 5°C} \\ -\rho_{40°C} + \frac{(45°C - 40°C)(\rho_{50°C} - \rho_{40°C})}{50°C - 40°C} \end{array}\right) \\
&= (250 \text{ mL})\left(\frac{1 \text{ L}}{1000 \text{ mL}}\right)\left(\frac{1 \text{ m}^3}{1000 \text{ L}}\right)\left(1000 \frac{\text{g}}{\text{kg}}\right) \\
&\quad \times \left(\begin{array}{c} 1000 \frac{\text{kg}}{\text{m}^3} + \frac{(6°C - 5°C)\left(999.7 \frac{\text{kg}}{\text{m}^3} - 1000 \frac{\text{kg}}{\text{m}^3}\right)}{10°C - 5°C} \\ -992.2 \frac{\text{kg}}{\text{m}^3} + \frac{(45°C - 40°C)\times\left(988 \frac{\text{kg}}{\text{m}^3} - 992.2 \frac{\text{kg}}{\text{m}^3}\right)}{50°C - 40°C} \end{array}\right) \\
&= 2.5 \text{ g}
\end{aligned}$$

**The answer is (C).**

**73.** The equation for the octanol-water partition coefficient is

**Octanol-Water Partition Coefficient**
$$K_{ow} = C_o/C_w$$

If $V_w$ is the volume of the water phase and $V_o$ is the volume of the octanol phase, the total mass of chlorpyrifos in both phases is

$$m = V_w C_w + V_o C_o$$

Rearrange to solve for $C_o$.

$$C_o = K_{ow}C_w$$
$$m = V_w C_w + V_o K_{ow} C_w$$
$$= C_w(V_w + V_o K_{ow})$$
$$C_w = \frac{m}{V_w + V_o K_{ow}}$$
$$C_o = \frac{K_{ow} m}{V_w + V_o K_{ow}}$$
$$= \frac{(4.7)(0.02 \text{ mg})}{100 \text{ L} + (4.7)(3 \text{ L})}$$
$$= 8.2 \times 10^{-4} \text{ mg/L}$$

**The answer is (A).**

**74.** The equation for the hazard index is

**Noncarcinogens**

$$HI = CDI_{\text{noncarcinogen}}/RfD$$

$CDI_{\text{noncarcinogen}}$ is the average daily intake of the compound, and $RfD$ is the reference dose. For methyl mercury ingestion, the hazard index is

$$HI = \frac{\left(1 \dfrac{\mu g}{kg \cdot d}\right)\left(\dfrac{1 \text{ mg}}{1000 \ \mu g}\right)}{0.0001 \dfrac{\text{mg}}{kg \cdot d}} = 10$$

**The answer is (C).**

**75.** Use an area-weighted average to determine the overall runoff coefficient for the watershed. $C$ is runoff coefficient and $A$ is area.

$$C_{\text{overall}} = \frac{C_{\text{concrete}} A_{\text{concrete}} + C_{\text{woods}} A_{\text{woods}}}{A_{\text{concrete}} + A_{\text{woods}}}$$
$$= \frac{(0.90)(18 \text{ ac}) + (0.05)(2 \text{ ac})}{18 \text{ ac} + 2 \text{ ac}}$$
$$= 0.82$$

**The answer is (B).**

**76.** This is a mass balance problem. The minimum necessary capacity is the capacity for which the water held to meet the demand never drops below zero. Assemble a table that shows the cumulative requirements for this reservoir, without allowing the reservoir capacity of the first season to be negative; then the second season, then the third season, and so on. Include more than one year in order to ensure that any carry-over need from a prior year is met. The maximum amount of water held in the reservoir is the minimum reservoir size. [**Mass Calculations**]

water held in reservoir
at end of this season = water held in reservoir
at end of previous season
+ inflow during this season
− demand during this season

| season | inflow | demand | water held in reservoir at end of season |
|---|---|---|---|
| fall$_1$ | 1 | 2 | 0 |
| winter$_1$ | 7 | 0 | 7 |
| spring$_1$ | 1 | 2 | 6 |
| summer$_1$ | 0 | 5 | 1 |
| fall$_2$ | 1 | 2 | 0 |
| winter$_2$ | 7 | 0 | 7 |
| spring$_2$ | 1 | 2 | 6 |
| summer$_2$ | 0 | 5 | 1 |

The minimum necessary capacity is 7 km$^3$.

**The answer is (C).**

**77.** This is a mass balance problem. This solution assumes a basis of pre-development stormwater loading, $L_{\text{pre}}$, of 1 kg/km$^2$, but any basis could be selected. The post-development loading without stormwater management strategies is 15 times that of the pre-development loading, and that after development the loading must not exceed pre-development loading. [**Mass Calculations**]

$$L_{\text{post,in}} = 15(L_{\text{pre}}) = (15)\left(1 \ \frac{\text{kg}}{\text{km}^2}\right) = 15 \text{ kg/km}^2$$

$$L_{\text{post,out}} = L_{\text{pre}} = 1 \text{ kg/km}^2$$

The removal efficiency can be calculated by applying the formula for the removal efficiency of incineration.

**Incineration**

$$\text{removal efficiency} = \left(\frac{L_{\text{post,in}} - L_{\text{post,out}}}{L_{\text{post,in}}}\right) \times 100\%$$
$$= \left(\frac{15 \ \dfrac{\text{kg}}{\text{km}^2} - 1 \ \dfrac{\text{kg}}{\text{km}^2}}{15 \ \dfrac{\text{kg}}{\text{km}^2}}\right) \times 100\%$$
$$= 93\%$$

**The answer is (D).**

**78.** Transmissivity for confined aquifers is related to hydraulic conductivity and aquifer depth as shown.

*Thiem Equation*
$$T = Kb$$

Rearrange and solve for the hydraulic conductivity.

$$K = \frac{T}{b}$$
$$= \frac{1300 \ \frac{\text{ft}^2}{\text{day}}}{45 \ \text{ft}}$$
$$= 29 \ \text{ft/day}$$

**The answer is (B).**

**79.** There are two log cycles that can be used with the graph shown: one that goes from 10 min to 100 min, and one that goes from 100 min to 1000 min. They both yield the same answer. In this solution, the log cycle that goes from 10 to 100 is used.

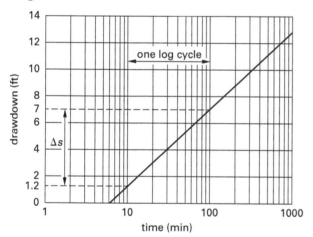

The value of the drawdown at 10 min is 1.2 ft, and the value of drawdown at 100 min is 7 ft.

$$\Delta s = 7 \ \text{ft} - 1.2 \ \text{ft} = 5.8 \ \text{ft}$$

**The answer is (D).**

**80.** Effective porosity is not needed to solve this problem. The Darcy velocity is

*Darcy's Law*
$$q = \frac{Q}{A} = -K(dh/dx)$$
$$= -\left(67 \ \frac{\text{m}}{\text{d}}\right)(-0.5\%)$$
$$= 0.33 \ \text{m/d}$$

**The answer is (A).**

**81.** The equation for calculating hydraulic conductivity from a constant head test is

*Geotechnical*
$$k = Q/(iAt_e)$$

The equation for $i$ is

$$i = dh/dL$$

Substitute and rearrange.

$$k = \left(\frac{Q}{t_e}\right)\left(\frac{dL}{A\,dh}\right)$$

$dL$ is the depth of the soil sample, and $Q/t_e$ is the flow rate through the permeameter. The cross-sectional area of the permeameter, $A$, can be calculated from the diameter, $D$.

The head causing flow through the soil sample, $dh$, is the head from the top of the permeameter to the bottom of the soil, or the height of the water above the soil sample plus the depth of the soil sample.

$$dh = 2.4 \ \text{m} + 1.9 \ \text{m} = 4.3 \ \text{m}$$

Substitute values into the equation for hydraulic conductivity and solve.

$$k = \left(\frac{Q}{t_e}\right)\left(\frac{dL}{A\,dh}\right) = \left(\frac{Q}{t_e}\right)\left(\frac{dL}{\left(\frac{\pi D^2}{4}\right)dh}\right)$$

$$= \left(0.02 \ \frac{\text{m}^3}{\text{h}}\right)\left(\frac{(1.9 \ \text{m})}{\left(\frac{\pi(0.37 \ \text{m})^2}{4}\right)(4.3 \ \text{m})}\right)$$

$$= 0.082 \ \text{m/h} \quad (0.08 \ \text{m/h})$$

**The answer is (C).**

**82.** The non-aqueous phase liquid (NAPL) may be found in any of these phases in the soil matrix: gaseous, aqueous, solid, or NAPL.

**The answer is (D).**

**83.** The equation for the concentration of a first-order irreversible reaction at time $t$ is

*First-Order Irreversible Reaction*
$$\ln(C_A/C_{A0}) = -kt$$

$C_{A0}$ is the initial concentration of microorganisms, and $C_A$ is the concentration of microorganisms at time $t$.

Integrating from the initial time to a time $t$,

$$t = \left(\frac{2.303}{k}\right)\log_{10}\left(\frac{C_{A0}}{C_A}\right)$$

For a time of 1 min and a concentration that is 50% of the initial concentration,

$$2.303\log_{10}\left(\frac{C_{A0}}{0.5\,C_{A0}}\right) = k(1\text{ min})$$

$$k = 2.303\log_{10}(2)$$
$$= 0.693 \text{ per min}$$

The time needed to bring the concentration of microorganisms down to 0.1% is

$$t = \left(\frac{2.303}{k}\right)\log_{10}\left(\frac{C_{A0}}{C_A}\right)$$
$$= \left(\frac{2.303}{0.693/\text{min}}\right)\log_{10}\left(\frac{C_{A0}}{(0.001)\,C_{A0}}\right)$$
$$= 10 \text{ min}$$

**The answer is (C).**

**84.** Under average baffling conditions, the baffling factor is 0.5. [Baffling Factors]

The theoretical detention time is

$$T = \frac{12}{C(\text{BF})}$$
$$= \frac{12}{\left(20\,\dfrac{\text{mg}}{\text{L}}\right)(1.8)(0.5)}$$
$$= 0.67 \text{ min}$$

**The answer is (B).**

**85.** The initial total water usage is (175 gpcd)$P_o$, where $P_o$ is the initial population. The increase in the water consumption rate is linear for 10 years at 0.1% per year. The rate at the end of 10 years is

$$\left(175\,\frac{\text{gal}}{\text{capita-day}}\right)\frac{(10\text{ yr})(0.1\%)}{100\%}$$
$$+ 175\,\frac{\text{gal}}{\text{capita-day}} = 176.50 \text{ gpcd}$$

The population increase is

$$P_{10} = P_o \exp(0.1)(10) = 2.718 P_o$$

The total water usage after 10 years is $(176.50)(2.718 P_o) = (479.72) P_o$. The percentage increase in water usage in 10 years is equal to

$$\frac{\left(479.72\,\dfrac{\text{gal}}{\text{capita-day}}\right)P_o - \left(175\,\dfrac{\text{gal}}{\text{capita-day}}\right)P_o}{\left(175\,\dfrac{\text{gal}}{\text{capita-day}}\right)P_o} \times 100\% = 174\% \quad (170\%)$$

**The answer is (D).**

**86.** The population after 10 years compounded at 1.1% per year is

$$(540{,}000\text{ people})(1+i)^n = (540{,}000\text{ people})(1.011)^{10}$$
$$= 602{,}428 \text{ people}$$

The per capita usage after 10 years is

$$\left(181\,\frac{\text{gal}}{\text{capita-day}}\right)(1.0002)^{10} = 181.36 \text{ gcpd}$$

The total water usage per day after 10 years is

$$\left(181.36\,\frac{\text{gal}}{\text{capita-day}}\right)$$
$$\times (602{,}428 \text{ people}) = 109{,}250{,}000 \text{ gpd}$$
$$(110{,}000{,}000 \text{ gpd})$$

**The answer is (B).**

**87.** The equation for the head loss in a pipe is

**Head Loss Due to Flow**

$$h_f = (4 f_{\text{Fanning}})\frac{L v^2}{D 2g}$$

$h =$ loss of head

$f =$ friction factor

$Q =$ volumetric flow rate, m³/s

$D =$ diameter, m

$L =$ length, m

$g =$ gravitational acceleration, 9.81 m/s²

The equation for the velocity in the pipe is

$$v = \frac{Q}{\pi\dfrac{D^2}{4}}$$

Substitute the equation for velocity into the equation for head loss.

$$h_f = (4f_{Fanning})\frac{Lv^2}{D2g}$$

$$= (4f_{Fanning})\frac{L\left(\dfrac{Q}{\pi\dfrac{D^2}{4}}\right)^2}{D2g}$$

$$= \frac{4f_{Fanning}LQ^2(4)^2}{(2g\pi^2 D)D^4}$$

$$= \frac{f_{Fanning}Q^2 L}{3D^5}$$

The two 500 m sections of pipe can be treated as two separate pipes with different diameters in series. The equation for the total head loss through the pipe is

$$h_1 = \frac{f_{Fanning,1}Q_1^2 L_1}{3D_1^5} + \frac{f_{Fanning,2}Q_2^2 L_2}{3D_2^5}$$

The flow rate is the same for both the pipe sections.

$$Q_1 = Q_2 = Q$$

The total flow rate is

$$Q = \sqrt{h_1\left(\frac{3D_1^5}{f_{Fanning,1}L_1} + \frac{3D_2^5}{f_{Fanning,2}L_2}\right)}$$

$$= \sqrt{(25\text{ m})\left(\frac{(3)(50\text{ cm})^5}{(0.0055)(500\text{ m})\left(100\dfrac{\text{cm}}{\text{m}}\right)^5} + \frac{(3)(30\text{ cm})^5}{(0.0055)(500\text{ m})\left(100\dfrac{\text{cm}}{\text{m}}\right)^5}\right)}$$

$$= 0.96\text{ m}^3/\text{s} \quad (1.00\text{ m}^3/\text{s})$$

**The answer is (A).**

**88.** The equation for volumetric flow through a pipe is

$$Q = vA = \frac{v\pi D^2}{4}$$

The Fanning friction factor and Darcy friction factor are related, as shown.

$$f_{Fanning} = 0.25 f_{Darcy}$$

Head loss from flow through a pipe is

$$\Delta h = \frac{2f_{Fanning}v^2 L}{gD} = \frac{0.5 f_{Darcy}v^2 LA^2}{gD\left(\dfrac{\pi D^2}{4}\right)^2} = \frac{8f_{Darcy}Q^2 L}{g\pi^2 D^5}$$

Solve for $Q$.

$$Q = \sqrt{\frac{\Delta h g\pi^2 D^5}{8f_{Darcy}L}}$$

The flow rate through the first pipe is

$$Q_1 = \sqrt{\frac{Hg\pi^2 D_1^5}{8f_{1,Darcy}L_1}} = \sqrt{\frac{(4\text{ m})(\pi^2)(9.8\text{ m/s}^2)(1\text{ m})^5}{8(0.1)(2500\text{ m})}}$$

$$= 0.44\text{ m}^3/\text{s}$$

The flow rate through the second pipe is

$$Q_2 = \sqrt{\frac{Hg\pi^2 D_2^5}{8f_{2,Darcy}L_2}} = \sqrt{\frac{(4\text{ m}(9.8\text{ m/s}^2)\pi^2(1.1\text{ m})^5}{8(0.072)(2498\text{ m})}}$$

$$= 0.66\text{ m}^3/\text{s}$$

The total flow between the tanks is

$$0.44\ \frac{\text{m}^3}{\text{s}} + 0.66\ \frac{\text{m}^3}{\text{s}} = 1.1\text{ m}^3/\text{s}$$

**The answer is (B).**

**89.** For a weir where the end contractions have been suppressed, the flow rate is given by the equation shown.

**Weir Formulas: Rectangular**

$$Q = CLH^{3/2}$$

$C = 1.84$ for SI units

$L$ = weir length, m

$H$ = depth of discharge over weir, m

The flow rate is

$$Q = (1.84)(10\text{ m})\left(\frac{50\text{ cm}}{100\dfrac{\text{cm}}{\text{m}}}\right)^{3/2}\left(60\ \frac{\text{s}}{\text{min}}\right)$$

$$= 390\text{ m}^3/\text{min} \quad (390\text{ m}^3/\text{min})$$

**The answer is (D).**

**90.** Express the number of moles of nitrogen in tank 1 as a function of the volume and the universal gas constant. [**Ideal Gas Constants**]

$$N_1 = \frac{(PV)_{\text{initial}_1}}{(RT)_1} = \frac{(1.2 \text{ atm})V}{R(20°C + 273°)}$$

$$= \frac{(1.2 \text{ atm})V}{R(293\text{K})}$$

$$= (V/R)(0.004095 \text{ atm/K})$$

Express the number of moles of nitrogen in tank 1 as a function of the volume and the universal gas constant.

$$N_2 = \frac{(PV)_{\text{initial}_2}}{(RT)_2}$$

$$= \frac{(0.8 \text{ atm})V}{R(25°C + 273°)}$$

$$= (V/R)(0.002684 \text{ atm/K})$$

Combine the two equations to find the total moles in the gas mixture created after the opening of the valve.

$$\frac{(1.2 \text{ atm})V}{R(293\text{K})} + \frac{(0.8 \text{ atm})V}{R(298\text{K})}$$

$$= \frac{V}{R}\left(0.004095 \frac{\text{atm}}{\text{K}} + 0.002684 \frac{\text{atm}}{\text{K}}\right)$$

$$= (V/R)(0.006779 \text{ atm/K})$$

The final temperature of the mixture, $T_f$, is the average of the mole-weighted averages of the contents of both tanks.

$$\left(0.004095 \frac{\text{atm}}{\text{K}}\right) \times (298\text{K} - T_f) = \left(0.002684 \frac{\text{atm}}{\text{K}}\right)(T_f - 293\text{K})$$

$$T_f = \frac{\left(0.004095 \frac{\text{atm}}{\text{K}}\right)(298\text{K}) + \left(0.002684 \frac{\text{atm}}{\text{K}}\right)(293\text{K})}{0.006779 \frac{\text{atm}}{\text{K}}} - 273°$$

$$= 23°C$$

**The answer is (C).**

**91.** For an ideal gas, the relationship between the specific heat at constant pressure, the specific heat at constant volume, and the specific gas constant is

**Ideal Gas Constants**

$$c_p - c_v = R$$

Therefore,

$$c_v = c_p - R$$
$$= 0.8RT - 0.1R$$

**The answer is (D).**

**92.** A life-cycle cost analysis determines the cost of a system over the course of its lifetime, from development to end-of-life. Because of the time value of money and the expected lifetime of transportation projects, discounting must be applied to future costs so that they can be combined with initial costs; for example, in a transportation project, future costs of maintenance and other future costs are discounted so that they can be combined with construction costs for a side-by-side comparison of alternatives. In a transportation project, users may incur higher costs due to more frequent maintenance-related closures for some alternatives than for others, and some transportation agencies take these user costs into account when conducting life-cycle cost analysis.

**The answer is (D).**

**93.** From a table of common names and molecular formulas of industrial chemicals, the molecular formula for vinyl chloride is $CH_2$=CHCl. [**Common Names and Molecular Formulas of Some Industrial (Inorganic and Organic) Chemicals**]

**The answer is (A).**

**94.** This molecule is a carboxylic acid with 10 straight-chain carbon atoms. Per the rules of nomenclature in organic chemistry, this is capric acid. [**Selected Rules of Nomenclature in Organic Chemistry**]

**The answer is (A).**

**95.** This is a mass balance problem. Perform a mass balance for water around the junction of the La Plata River with the San Juan River (point A) to determine the flow rate of the San Juan River at point A. $Q$ is volumetric flow rate. [**Mass Calculations**]

$$Q_A = Q_B + Q_D$$
$$= 1673 \frac{\text{ft}^3}{\text{sec}} + 2 \frac{\text{ft}^3}{\text{sec}}$$
$$= 1675 \text{ ft}^3/\text{sec}$$

A mass balance for arsenic from point A to point C yields the equation shown. $\dot{m}$ is the mass flow rate of arsenic.

$$\dot{m}_A = \dot{m}_C + \dot{m}_D + \dot{m}_E = \dot{m}_C$$

If $C$ is concentration in terms of mass per unit volume, the fate and transport mass calculations can be represented as shown.

$$Q_A C_A = Q_C C_C$$

Solve for $C_C$.

$$C_C = \frac{Q_A C_A}{Q_C}$$

$$= \frac{\left(1675 \, \frac{\text{ft}^3}{\text{sec}}\right)\left(11 \, \frac{\mu\text{g}}{\text{L}}\right)}{776 \, \frac{\text{ft}^3}{\text{sec}}}$$

$$= 23 \, \mu\text{g/L}$$

Note: The flow rate of the La Plata River is so small it can be neglected.

**The answer is (C).**

**96.** The $LD_{50}$ is the median lethal single dose, the single dose at which half of the population dies. Calculate the fraction of the population that died at each dose, which is the number of rats that died divided by the number of rats dosed. [**Dose-Response Curves**]

| dose (mg/kg of body weight) | fraction of rats that died |
|---|---|
| 100 | 0 |
| 500 | 0.1 |
| 1000 | 0.5 |
| 1500 | 0.7 |
| 2000 | 1.0 |

Half the population died at a dose of 1000 mg/kg of body weight.

**The answer is (B).**

**97.** The workroom is unvented, so the partial pressure of toluene in the room is equal to the vapor pressure. The partial pressure can be converted to parts per million using the ideal gas law and dividing the molar concentration of toluene by the molar concentration of air. The ideal gas law is

$$PV = nRT$$

$P$ is pressure, $V$ is volume, $n$ is the number of moles, $R$ is the ideal gas law constant, and $T$ is absolute temperature. Atmospheric pressure is 101.325 kPa.

$$\frac{\dfrac{n_{\text{toluene}}}{V_{\text{room}}}}{\dfrac{n_{\text{air}}}{V_{\text{room}}}} = \frac{\dfrac{P_{\text{vap,toluene}}}{RT}}{\dfrac{P_{\text{atm}}}{RT}} = \frac{n_{\text{toluene}}}{n_{\text{air}}}$$

$$= \frac{P_{\text{vap,toluene}}}{P_{\text{atm}}}$$

$$= \frac{3.8 \text{ kPa}}{101.325 \text{ kPa}}$$

$$= \left(0.0375 \, \frac{\text{mol toluene}}{\text{mol air}}\right)$$

$$\times \left(1 \, \frac{\text{mol air}}{\text{part air}}\right)\left(1 \, \frac{\text{part toluene}}{\text{mol toluene}}\right)$$

$$\times \left(\frac{1 \text{ ppm toluene}}{\left(\dfrac{1 \text{ part toluene}}{1{,}000{,}000 \text{ parts air}}\right)}\right)$$

$$= 37{,}500 \text{ ppm}$$

The permissible 8 hr time weighted average for toluene is 200 ppm, so this concentration is

$$\left(\frac{37{,}500 \text{ ppm}}{200 \text{ ppm}}\right) \times 100\% = \begin{array}{l} 19{,}000\% \text{ of the permissible} \\ \text{exposure level} \end{array}$$

**The answer is (C).**

**98.** Reference dose ($RfD$) is a metric for non-carcinogenic effects, and is the dose that a healthy person can be exposed to daily without adverse effects. The higher the reference dose, the less toxic a compound is. [**Reference Dose**]

**The answer is (A).**

**99.** Use the EPA-recommended values for estimating intake. For a female adult, the daily amount of water ingested is 2.3 L/d, and a lifetime is 75 yr. The exposure frequency is 365 d/yr. [**Intake Rates—Variable Values**]

The lifetime ingestion is

ingestion over lifetime = (number of years per lifetime)
$\times$(CW)(IR)(EF)

$$= \left(75 \frac{\text{yr}}{\text{lifetime}}\right)\left(0.007 \frac{\text{mg}}{\text{L}}\right)$$
$$\times \left(2.3 \frac{\text{L}}{\text{d}}\right)\left(365 \frac{\text{d}}{\text{yr}}\right)$$
$$= 440 \text{ mg/lifetime}$$

**The answer is (C).**

**100.** Use the Natural Resources Conservation Service (NRCS (SCS)) rainfall-runoff equation.

**NRCS (SCS) Rainfall-Runoff**

$$Q = \frac{(P - 0.2S)^2}{P + 0.8S}$$

Check to see if the initial abstraction is greater than the rainfall (in which case runoff is zero). The initial abstraction for the pre-development scenario is

$$0.2S = (0.2)\left(\frac{1000}{CA} - 10\right)$$
$$= (0.2)\left(\frac{1000}{74} - 10\right)$$
$$= 0.70 \text{ in}$$

For the post-development scenario, the initial abstraction is

$$0.2S = (0.2)\left(\frac{1000}{CA} - 10\right)$$
$$= (0.2)\left(\frac{1000}{90} - 10\right)$$
$$= 0.22 \text{ in}$$

In both cases the initial abstraction is less than the rainfall, so runoff will be more than 0 in for both scenarios.

Substitute the equation for the maximum basin retention into the NRCS (SCS) rainfall-runoff equation.

$$Q = \frac{\left(P - 0.2\left(\frac{1000}{CN} - 10\right)\right)^2}{P + 0.8\left(\frac{1000}{CN} - 10\right)}$$

Before development, the runoff is

$$Q = \frac{\left(P - 0.2\left(\frac{1000}{CN} - 10\right)\right)^2}{P + 0.8\left(\frac{1000}{CN} - 10\right)}$$

$$= \frac{\left(1 \text{ in} - (0.2)\left(\frac{1000}{74} - 10\right)\right)^2}{1 \text{ in} + (0.8)\left(\frac{1000}{74} - 10\right)}$$

$$= 0.023 \text{ in}$$

After development, runoff is

$$Q = \frac{\left(P - 0.2\left(\frac{1000}{CN} - 10\right)\right)^2}{P + 0.8\left(\frac{1000}{CN} - 10\right)}$$

$$= \frac{\left(1 \text{ in} - (0.2)\left(\frac{1000}{90} - 10\right)\right)^2}{1 \text{ in} + (0.8)\left(\frac{1000}{90} - 10\right)}$$

$$= 0.32 \text{ in}$$

After development, the runoff increases by a factor of

$$\frac{0.32 \text{ in}}{0.023 \text{ in}} = 14$$

**The answer is (D).**

**101.** This is a mass balance problem. The minimum necessary capacity is the capacity for which the water held to meet the demand never drops below zero. Tabulate the cumulative requirements for this reservoir without allowing the reservoir capacity of the first month to be negative, then the second month, then the third month, and so on. Include more than one year in order to ensure that any carry-over need from a prior year is met. The maximum amount of water required from the reservoir is the reservoir size that just meets demand. [Mass Calculations]

water held in reservoir at end of this season = water held in reservoir at end of previous season
+inflow during current season
−demand during current season

| month | inflow (km³) | demand (km³) | water held in reservoir at end of season (km³) |
|---|---|---|---|
| Jan, year 1 | 1 | 5 | 0 |
| Feb, year 1 | 2 | 3 | 0 |
| Mar, year 1 | 4 | 1 | 3 |
| Apr, year 1 | 7 | 0 | 10 |
| May, year 1 | 11 | 0 | 21 |
| Jun, year 1 | 6 | 0 | 27 |
| Jul, year 1 | 2 | 1 | 28 |
| Aug, year 1 | 2 | 2 | 28 |
| Sep, year 1 | 1 | 4 | 25 |
| Oct, year 1 | 0 | 5 | 20 |
| Nov, year 1 | 0 | 9 | 11 |
| Dec, year 1 | 0 | 6 | 5 |
| Jan, year 2 | 1 | 5 | 1 |
| Feb, year 2 | 2 | 3 | 0 |
| Mar, year 2 | 4 | 1 | 3 |
| Apr, year 2 | 7 | 0 | 10 |
| May, year 2 | 11 | 0 | 21 |
| Jun, year 2 | 6 | 0 | 27 |
| Jul, year 2 | 2 | 1 | 28 |
| Aug, year 2 | 2 | 2 | 28 |
| Sep, year 2 | 1 | 4 | 25 |
| Oct, year 2 | 0 | 5 | 20 |
| Nov, year 2 | 0 | 9 | 11 |
| Dec, year 2 | 0 | 6 | 5 |

The minimum necessary capacity is 28 km³.

**The answer is (C).**

**102.** This is a mass (volume) balance problem. The equation for the water held in the reservoir at the end of a given hour is

**Mass Calculations**

water held in reservoir at end of this hour = water held in reservoir at end of previous hour
+ inflow during current hour
− outflow during current hour

Calculate the water held in the reservoir at the end of each time interval.

| time interval (h) | inflow (m³/h) | outflow (m³/h) | water held at end of time interval (m³) |
|---|---|---|---|
| 1 | 2000 | 0 | 2000 |
| 2 | 9000 | 2000 | 9000 |
| 3 | 10,000 | 7000 | 12,000 |
| 4 | 7000 | 8000 | 11,000 |
| 5 | 5000 | 8000 | 8000 |
| 6 | 3000 | 5000 | 6000 |
| 7 | 2000 | 4000 | 4000 |
| 8 | 1000 | 3000 | 2000 |
| 9 | 1000 | 2000 | 1000 |
| 10 | 1000 | 2000 | 0 |
| 11 | 1000 | 1000 | 0 |
| 12 | 1000 | 1000 | 0 |

The peak water held is 12,000 m³.

Note: It is not necessary to calculate the values beyond the first three hours because after that the outflow exceeds the inflow.

**The answer is (C).**

**103.** Variable assignments for the streams and processes are as shown.

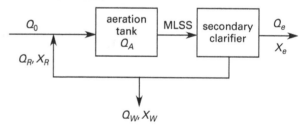

A mass flow balance around the entire system gives [Mass Calculations]

$$Q_0 = Q_e + Q_W$$

Rearrange to solve for the effluent flow rate.

$$Q_e = Q_0 - Q_W$$

The equation for the recycle ratio is

**Activated Sludge**

$$R = Q_R/Q_0$$

Rearrange to solve for the recycle flow rate.

$$Q_R = RQ_0$$

From the information given, the equation for effluent suspended solids concentration is

$$X_e = 0.1 X_A$$

Rearrange to solve for the suspended solids concentration in the aeration tank, $X_A$.

$$X_A = \frac{X_e}{0.1} = 10 X_e$$

The formula for the steady-state mass balance around a secondary clarifier is

**Activated Sludge**

$$(Q_0 + Q_R) X_A = Q_e X_e + Q_R X_r + Q_w X_w$$

Substitute in the relationships given in the problem statement and from the mass flow balance around the entire system.

$$(Q_0 + R Q_0) 10 X_e = (Q_0 - Q_W) X_e + R Q_0 X_r + Q_w X_w$$

Simplify, rearrange to solve for $X_e$, and substitute values.

$$(10 Q_0 + 10 R Q_0) X_e - (Q_0 - Q_W) X_e = R Q_0 X_w + Q_w X_w$$

$$X_e = \frac{R Q_0 X_w + Q_w X_w}{(10 Q_0 + 10 R Q_0) - (Q_0 - Q_W)}$$

$$= \frac{R Q_0 X_w + Q_w X_w}{Q_0(9 + 10R) + Q_W}$$

$$= \frac{(0.25)\left(440{,}000 \ \frac{\text{gal}}{\text{day}}\right)\left(10{,}000 \ \frac{\text{mg}}{\text{L}}\right) + \left(150{,}000 \ \frac{\text{gal}}{\text{day}}\right)\left(10{,}000 \ \frac{\text{mg}}{\text{L}}\right)}{\left(440{,}000 \ \frac{\text{gal}}{\text{day}}\right)(9 + (10)(0.25)) + 150{,}000 \ \frac{\text{gal}}{\text{day}}}$$

$$= 500 \text{ mg/L}$$

**The answer is (B).**

**104.** This is a mass balance problem. The outlet concentration of suspended solids from the aeration tank is the concentration of suspended solids in the aeration tank.

**Mass Calculations**

**Activated Sludge**

$$(\text{mass suspended solids})_{\text{in}} = (\text{mass suspended solids})_{\text{out}}$$

$$Q_0 X_0 + Q_R X_r = (Q_0 + Q_R) X_A$$

Rearrange to solve for the flow rate of the returned sludge and substitute values.

$$Q_0 X_0 + Q_R X_r = Q_0 X_A + Q_R X_A$$
$$Q_R(X_r - X_A) = Q_0(X_A - X_0)$$
$$Q_R = \frac{Q_0(X_A - X_0)}{X_r - X_A}$$

$$= \frac{\left(3{,}600{,}000 \ \frac{\text{gal}}{\text{day}}\right)\left(2500 \ \frac{\text{mg}}{\text{L}} - 2000 \ \frac{\text{mg}}{\text{L}}\right)}{9000 \ \frac{\text{mg}}{\text{L}} - 2500 \ \frac{\text{mg}}{\text{L}}}$$

$$= 280{,}000 \text{ gpd}$$

**The answer is (C).**

**105.** Combine the formula for calculating the volume of a right circular cylinder with the factor for converting cubic feet to gallons. The volume is

**Right Circular Cylinder**

**Conversion Factors**

$$V = \frac{\pi d^2 h}{4}$$

$$= \left(\frac{\pi (3.0 \text{ in})^2 \left(\frac{1 \text{ ft}}{12 \text{ in}}\right)^2 (200 \text{ ft})}{4}\right)\left(\frac{1 \text{ gal}}{0.134 \text{ ft}^3}\right)$$

$$= 73 \text{ gal}$$

**The answer is (B).**

**106.** The design equation for biotowers is

**Biotower**

$$\frac{S_e}{S_0} = e^{-kD/q^n}$$

The treatability constant is 0.06 min$^{-1}$ at 20°C, and the coefficient relating to media characteristics for modular plastic media is 0.5. [Biotower]

The treatability constant for this biotower must be converted to the treatability at 15°C using the following equation.

**Biotower**

$$k_T = k_{20}(1.035)^{T-20}$$

The hydraulic loading rate, $q$, for a non-recycling biotower is the flow rate, $Q_0$, divided by the cross-sectional area of the tower, $A_{\text{plan}}$.

$$q = \frac{Q_0}{A_{\text{plan}}}$$

The equation for the cross-sectional area of a biotower with diameter $d$ is

$$A_{\text{plan}} = \frac{\pi d^2}{4}$$

The equation for the hydraulic loading rate becomes

$$q = \frac{4Q_0}{\pi d^2}$$

Rearrange the design equation to solve for the depth of the biotower media, $D$, and substitute values. Note that $q$ must be in units of m/min.

$$\ln\left(\frac{S_e}{S_0}\right) = \frac{-kD}{q^n}$$

$$D = -\frac{q^n}{k}\ln\left(\frac{S_e}{S_0}\right)$$

$$= -\frac{\left(\dfrac{4Q_0}{\pi d^2}\right)^n}{k_{20}(1.035)^{T-20}}\ln\left(\frac{S_e}{S_0}\right)$$

$$= -\frac{\left(\dfrac{(4)\left(450\ \dfrac{\text{m}^3}{\text{d}}\right)\left(\dfrac{1\ \text{d}}{1440\ \text{min}}\right)}{\pi(5\ \text{m})^2}\right)^{0.5}}{\left(0.06\ \dfrac{1}{\text{min}}\right)(1.035)^{15°C-20°C}}\ln\left(\dfrac{20\ \dfrac{\text{mg}}{\text{L}}}{180\ \dfrac{\text{mg}}{\text{L}}}\right)$$

$$= 5.5\ \text{m}$$

**The answer is (D).**

**107.** The equation for calculating hydraulic conductivity from a falling head test is

**Geotechnical**

$$k = 2.303[(aL)/(At_e)]\log(h_1/h_2)$$

Per the problem statement, $a = A$. The hydraulic conductivity of the soil is

$$k = \frac{2.303\,aL}{At_e}\log\left(\frac{h_1}{h_2}\right) = \frac{2.303L}{t_e}\log\left(\frac{h_1}{h_2}\right)$$

$$= \left(\frac{(2.303)(0.27\ \text{m})}{\left((2\ \text{h})\left(60\ \dfrac{\text{min}}{\text{h}}\right) + 37\ \text{min}\right)\left(60\ \dfrac{\text{s}}{\text{min}}\right)}\right)\log\left(\frac{0.41\ \text{m}}{0.023\ \text{m}}\right)$$

$$= 8.3 \times 10^{-5}\ \text{m/s}$$

**The answer is (A).**

**108.** The equation for the degree of saturation is

**Phase Relationships**

$$S = (V_W/V_V) \times 100\%$$

At saturation, $S = 100\%$. Substitute and solve for $V_W$ in terms of $V_V$.

$$100\% = \frac{V_W}{V_V} \times 100\%$$

$$V_W = V_V \frac{100\%}{100\%} = V_V$$

Options C and D are incorrect because they are always true, not just at saturation. Option B is incorrect because when that condition is met the soil is comprised of soil solids only, with no voids.

**The answer is (A).**

**109.** The equation for the power consumed is

**Blowers**

$$P_W = \frac{WRT_1}{Cne}\left[\left(\frac{P_2}{P_1}\right)^{0.283} - 1\right]$$

$R$ is the engineering gas constant for air. The mass flow rate of air is required, so volumetric flow rate has to be converted to mass flow rate. From the ideal gas law at these conditions,

$$W = \frac{m\dot{V}}{V} = \frac{\dot{V}MW_{\text{air}}n}{V}$$

$$= \frac{\dot{V}MW_{\text{air}}P}{RT}$$

$$= \frac{\left(28.9\ \dfrac{\text{lbm}}{\text{lbm-mol}}\right)(1\ \text{atm})}{\left(0.7302\ \dfrac{\text{ft}^3\text{-atm}}{\text{lbm-mol°R}}\right)\left((25°C)\left(\dfrac{9°R}{5°C}\right) + 491°R\right)}$$

$$= (0.074\ \text{lbm/ft}^3)\,\dot{V}$$

The power consumed is

$$P_W = \frac{WRT_1}{Cne}\left[\left(\frac{P_2}{P_1}\right)^{0.283} - 1\right]$$

$$= \frac{\left(0.074\ \frac{\text{lbm}}{\text{ft}^3}\right)\dot{V}RT_1}{Cne}\left[\left(\frac{P_2}{P_1}\right)^{0.283} - 1\right]$$

$$= \left[\frac{\left(0.074\ \frac{\text{lbm}}{\text{ft}^3}\right)\left(3000\ \frac{\text{ft}^3}{\text{min}}\right)\left(\frac{1\ \text{min}}{60\ \text{sec}}\right)}{\left(550\ \frac{\text{ft-lbm}}{\text{sec-hp}}\right)(0.283)(68\%)}\right]$$

$$\times \left[\left(\frac{2.5\ \text{atm}}{1\ \text{atm}}\right)^{0.283} - 1\right]$$

$$= 300\ \text{hp}$$

**The answer is (C).**

# Practice Exam 1

## PROBLEMS

**1.** A sharp pipe (diameter 4 cm) exits a large tank. The head loss at the entrance to the pipe, for the case of water at 25°C flowing at the rate of 1.0 L/min, is most nearly

(A)  $3.7 \times 10^{-6}$ m

(B)  $4.0 \times 10^{-6}$ m

(C)  $4.5 \times 10^{-6}$ m

(D)  $5.0 \times 10^{-6}$ m

**2.** The equilibrium equation for the dissolution of sulfur dioxide in water is

$$SO_2(g) + 2H_2O(l) \rightleftharpoons HSO_3^-(aq) + H_3O^+(aq)$$

The equilibrium constant for this equation is

$$K_{eq} = \frac{[HSO_3^-][H_3O^+]}{p_{SO_2}} = 2.1 \times 10^{-2} \ (mol/L)^2/atm$$

The variable $p_{SO_2}$ represents the partial pressure of sulfur dioxide in the gaseous phase. The molar concentration of hydronium ions in water in equilibrium with air that contains 0.1 parts per million volume (ppmv) of sulfur dioxide is most nearly

(A)  $2.1 \times 10^{-9}$ mol/L

(B)  $4.6 \times 10^{-5}$ mol/L

(C)  $5.3 \times 10^{-5}$ mol/L

(D)  $2.1 \times 10^{-3}$ mol/L

**3.** Alpha particles are stopped by

(A)  concrete, but not lead

(B)  lead and concrete, but not aluminum

(C)  aluminum, lead, and concrete, but not skin

(D)  aluminum, lead, concrete, and skin

**4.** The given venturi meter in water service has a coefficient of velocity of 0.85.

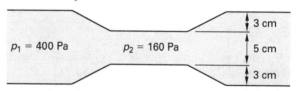

The specific weight, $\gamma$, of water is 9.8 kN/m³. The flow rate is most nearly

(A)  $6.3 \times 10^{-4}$ m³/s

(B)  $5.7 \times 10^{-4}$ m³/s

(C)  $1.2 \times 10^{-3}$ m³/s

(D)  $5.7 \times 10^{-1}$ m³/s

**5.** The parallel pipes shown carry water at 20°C with the same velocity.

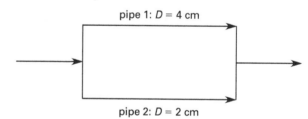

The ratio of power lost in pipe 1 to power lost in pipe 2 is

(A)  0.5

(B)  2.0

(C)  4.0

(D)  16

**6.** The ultimate biological oxygen demand (BOD), $S_0$, in a stream immediately following a wastewater discharge is 7 mg/L. The initial dissolved oxygen deficit, $D_a$, is 0 mg/L. The deoxygenation rate constant, $k_d$, is 0.1/d, and the reaeration rate of the stream, $k_r$, is 0.35/d. The number of days the stream must travel to reach a dissolved oxygen deficit, $D$, of 0.15 mg/L is most nearly

(A) 10 d
(B) 20 d
(C) 30 d
(D) 40 d

**7.** A centrifugal pump with a 1 m diameter impeller operates at 1000 rpm and generates a flow of 10 L/s. A second, similar pump operates at 1000 rpm, generating double the flow, or 20 L/s. The diameter of the second pump's impeller would most nearly be

(A) 0.50 m
(B) 1.0 m
(C) 1.3 m
(D) 4.0 m

**8.** A centrifugal pump with a 1 m impeller diameter and rotating at 100 rpm with a discharge of 1 L/s consumes 10 kW. If the same system needs a pump with the same rotational speed but a power consumption of 20 kW, the impeller diameter selected would most nearly be

(A) 1.2 m
(B) 1.5 m
(C) 1.8 m
(D) 2.0 m

**9.** Terrain-adjusted stack height (TESH) is equal to the effective stack height $H$ (physical stack height, $h$, plus plume rise, $\Delta h$) less the maximum rise in terrain within 5 km of the stack, $T_r$. The plume rise for a 30 m stack with an exhaust temperature of 550K and a flow rate of 4.5 m³/s is 10 m. It has been determined that the TESH can be no less than 28 m in order to meet regulatory requirements. Therefore, the maximum rise in terrain within 5 km of the stack is most nearly

(A) 10 m
(B) 12 m
(C) 48 m
(D) 68 m

**10.** The sewer system for a newly planned city of population 20,000 is being designed. The ratio of the minimum flow rate to the average flow rate is most nearly

(A) 0.35
(B) 0.45
(C) 0.55
(D) 0.65

**11.** Lime scrubbing is used to neutralize the sulfur dioxide in a stack. The chemical equations are

$$CaO \text{ (lime)} + H_2O \rightarrow Ca(OH)_2$$
$$SO_2 + Ca(OH)_2 \rightarrow CaSO_3 + H_2O$$

The amount of lime required to neutralize each kilogram of sulfur dioxide is most nearly

(A) 0.88 kg CaO/kg $SO_2$
(B) 1.0 kg CaO/kg $SO_2$
(C) 1.1 kg CaO/kg $SO_2$
(D) 1.5 kg CaO/kg $SO_2$

**12.** The heats of formation, $\Delta H_f$, of liquid benzene, carbon dioxide, and liquid water are 11.7 kcal/mol, −94.1 kcal/mol, and −68.3 kcal/mol, respectively. The higher heating value (HHV) of benzene, estimated using the equation for combustion, is most nearly

(A) −780 kcal/mol
(B) −760 kcal/mol
(C) −170 kcal/mol
(D) 780 kcal/mol

**13.** A circular sewer line of diameter 40 cm has a full-flowing flow rate of 200 L/s. The flow rate when the depth of flow is 25 cm is most nearly

(A) 100 L/s
(B) 110 L/s
(C) 120 L/s
(D) 150 L/s

**14.** In rain, formaldehyde undergoes hydrolysis to form methylene glycol according to the equation

$$HCHO(aq) + H_2O(l) H_2C(OH)_2(aq)$$

The equilibrium constant for this equation is 2000. The concentration of methylene glycol in rain that has formadehyde present at 1 ppb (mass) is most nearly

(A) 60 ppb (mass)

(B) 1000 ppb (mass)

(C) 2000 ppb (mass)

(D) 3000 ppb (mass)

**15.** A compound in the atmosphere has a residence time of 10 yr, and its concentration in the atmosphere is steady at 2 ppm (volume). The total mass of the atmosphere is $5.2 \times 10^{18}$ kg. The molecular weight of the compound is 16 g/mol, and the molecular weight of air is 29 g/mol. The emissions of this compound to the atmosphere are most nearly

(A) $20 \times 10^6$ g/s

(B) $30 \times 10^6$ g/s

(C) $60 \times 10^6$ g/s

(D) $80 \times 10^6$ g/s

**16.** Henry's law constant for dimethyl sulfide is 7.1 L·atm/mol. The mole fraction of dimethyl sulfide in water that is in equilibrium with a concentration of $10^{-6}$ parts per billion volume (ppbv) dimethyl sulfide in air is most nearly

(A) $1.4 \times 10^{-16}$

(B) $7.1 \times 10^{-15}$

(C) $1.0 \times 10^{-14}$

(D) $2.3 \times 10^{-14}$

**17.** A circular sewer has a diameter of 50 cm. The flow rate for full flow is 50 L/s. For a flow rate of 20 L/s, the expected depth of flow is most nearly

(A) 18 cm

(B) 20 cm

(C) 24 cm

(D) 30 cm

**18.** The concentration of dissolved oxygen in a stream is 9.0 mg/L. The mole fraction of oxygen in this stream is most nearly

(A) $5.0 \times 10^{-6}$

(B) $9.0 \times 10^{-6}$

(C) $5.0 \times 10^{-3}$

(D) $9.0 \times 10^{-3}$

**19.** An off-gas contains 12 ppm (volume) hydrogen sulfide ($H_2S$). The flow rate of the off-gas is 0.5 m$^3$/s. The mass flow rate of hydrogen sulfide in the off-gas at standard temperature and pressure is most nearly

(A) $6.0 \times 10^{-6}$ mg/s

(B) $9.1 \times 10^{-6}$ mg/s

(C) 6.0 mg/s

(D) 9.1 mg/s

**20.** A city had a population of 100,000 in 1970, 150,000 in 1980, and 220,000 in 1990. The daily per-capita water consumption was 1.1 kg/d in 1970, 1.20 kg/d in 1980, and 1.31 kg/d in 1990. The total daily water consumption for the city in 2010 is expected to be most nearly

(A) $3.8 \times 10^5$ kg/d

(B) $4.8 \times 10^5$ kg/d

(C) $5.2 \times 10^5$ kg/d

(D) $7.3 \times 10^5$ kg/d

**21.** A solvent that is used in a labeling process is released to a room at a constant rate of 0.5 kg/h. The maximum permissible concentration of the solvent in air, under occupational safety and health regulations, is 1 mg/m$^3$. The air in the room is well mixed. The ventilation flow rate of uncontaminated air into the room required to meet the regulated concentration limit is most nearly

(A) 90 m$^3$/s

(B) 140 m$^3$/s

(C) 380 m$^3$/s

(D) 1800 m$^3$/s

**22.** The activity of a radionuclide was originally 500 Bq. The radionuclide has a half-life of 15 d. The activity of the radionuclide after 28 d is most nearly

(A) 77 Bq

(B) 140 Bq

(C) 290 Bq

(D) 340 Bq

**23.** A nonvolatile pollutant is degraded by natural processes when released to surface water. When a quantity of this pollutant is discharged to a pond, its concentration decays according to the equation

$$C(t) = C_0 e^{(-0.15\ 1/d)t}$$

$C_0$ is the initial concentration of the pollutant. The time it will take for the concentration of the pollutant to drop by three-quarters is most nearly

(A) 0.21 d

(B) 1.9 d

(C) 5.0 d

(D) 9.2 d

**24.** The concentration of a chemical species in a sample is measured to be 1000 ppm. The error in the measurement is $\pm 100$ ppm. The approximate error in the square of the concentration is

(A) 10 000 ppm²

(B) 20 000 ppm²

(C) 100 000 ppm²

(D) 200 000 ppm²

**25.** The flow rate of chlorine from a 30 m stack is 0.8 kg/s. The wind speed is 4 m/s, and the atmosphere is slightly unstable. The plume rise is 10 m. The ground-level centerline concentration 2 km downwind from the source is most nearly

(A) 1 mg/m³

(B) 2 mg/m³

(C) 4 mg/m³

(D) 8 mg/m³

**26.** The chemical oxygen demand (COD) of $C_4H_5NO_2$ is most nearly

(A) 0.33 g $O_2$/g $C_4H_5NO_2$

(B) 0.95 g $O_2$/g $C_4H_5NO_2$

(C) 1.1 g $O_2$/g $C_4H_5NO_2$

(D) 3.5 g $O_2$/g $C_4H_5NO_2$

**27.** What is the most suitable material for a storage container meant to store concentrated nitric acid?

(A) tin

(B) steel

(C) alloys of aluminum or chromium

(D) tungsten

**28.** Two municipal waste streams are to be composted: grass clippings with a carbon:nitrogen mass ratio of 15, and brushy yard trimmings with a carbon:nitrogen ratio of 55. The ideal carbon:nitrogen mass ratio for composting is 30. The ideal mass percentage of grass clippings in the mixed stream is most nearly

(A) 37%

(B) 45%

(C) 63%

(D) 73%

**29.** The half-life of a biologically degraded, nonvolatile surface-water contaminant is 3 wk. In an accidental spill, 100 kg of the contaminant are released to a pond. After 4 wk, the amount of contaminant remaining in the pond is most nearly

(A) 10 kg

(B) 20 kg

(C) 30 kg

(D) 40 kg

**30.** The saturation dissolved oxygen concentration of a stream is 9.1 mg/L. At a sewage outfall, the dissolved oxygen concentration of the stream is 8.0 mg/L. The stream has a reaeration rate constant of 4/d and a deoxygenation rate constant of 0.1/d. The initial BOD ultimate in the mixing zone is 200 mg/L. The time after discharge at which the water will reach its minimum dissolved oxygen concentration is most nearly

(A) 1.0 h

(B) 1.7 h

(C) 21 h

(D) 34 h

**31.** The surface area of a reservoir, in square meters, depends on the depth of the water in the reservoir, $h$, according to the expression

$$A = 3 \times 10^6 \text{ m}^2 + (2 \times 10^4 \text{ m})h$$

Water is flowing into the reservoir at a rate of 50 m³/s. There are no flows out of the reservoir, and evaporation can be neglected. The depth of the water in the lake is 10 m. The amount of time it will take for the reservoir's depth to reach 11 m is most nearly

(A) 0.48 d
(B) 0.74 d
(C) 1.5 d
(D) 3.2 d

**32.** The toxicity characteristic leaching procedure (TCLP)

(A) identifies the degree to which a toxic compound passes through human skin
(B) quantifies the potential of a toxic material to leach through metal drums
(C) models the process that causes toxic compounds to be extracted from their original matrix into landfill leachate
(D) identifies chemicals that are mutagenic, teratogenic, or carcinogenic

**33.** At 25°C, the acid ionization constant for acetic acid is $1.77 \times 10^{-5}$ and the autoionization constant for water is $1 \times 10^{-14}$. The base ionization constant of acetic acid at 25°C is most nearly

(A) $1.8 \times 10^{-19}$
(B) $4.5 \times 10^{-10}$
(C) $5.6 \times 10^{-10}$
(D) $1.8 \times 10^{-9}$

**34.** The compression of baling of solid waste prior to disposal in a sanitary landfill

(A) reduces the chances that water will be polluted by leachates
(B) provides for material recovery
(C) provides for energy recovery
(D) conserves landfill space because the volume of solid waste is reduced

**35.** The potential for landfill leachates to contaminate groundwater increases with

(A) decreasing distance of the landfill from the nearest point-of-use water source
(B) increasing permeability of the soil where the landfill is located
(C) increasing distance from the landfill to the nearest aquifer
(D) decreasing rate of precipitation

**36.** A fluid is being ultrafiltered with a membrane that has a porosity of 50%, a pore size of 2 μm, and a thickness of 100 μm. For a fluid viscosity of 0.5 cP across a 10 000 Pa pressure drop, the volumetric flux during ultrafiltration is most nearly

(A) 0.0001 m/s
(B) 0.001 m/s
(C) 0.005 m/s
(D) 0.05 m/s

**37.** A horizontal rectangular stilling basin 20 m wide has a design discharge of 1000 m³/s. A detailed design is to be done using a scale model (1:20). The detailed design should be done using which dimensionless number(s)?

I. Reynolds number
II. Froude number
III. Weber number

(A) I
(B) II
(C) I and II
(D) I and III

**38.** A soil sample in a permeameter has a length of 0.5 ft and a cross-sectional area of 0.07 ft². The water head is 3.5 ft upstream and 0.1 ft downstream. A flow rate of 1 ft³/day is observed. The hydraulic conductivity is most nearly

(A) $2.4 \times 10^{-5}$ ft/sec
(B) $6.1 \times 10^{-5}$ ft/sec
(C) $8.2 \times 10^{-5}$ ft/sec
(D) $9.3 \times 10^{-5}$ ft/sec

**39.** Stormwater flows through a square concrete pipe that has a hydraulic radius of 1 m and an energy grade line slope of 0.8. Using the Hazen-Williams equation, the velocity of the water is most nearly

(A) 43 m/s
(B) 67 m/s
(C) 98 m/s
(D) 150 m/s

**40.** The reactor volume of a standard-rate anaerobic digester needs to be estimated for a given treatment process. The raw sludge input rate is 350 ft$^3$/day. The digested sludge accumulation is 1 ft$^3$/day. If the storage time is 12 hr and the time to react in a high-rate digester is 8 hr, the reactor volume required would be most nearly

(A) 18 ft$^3$
(B) 59 ft$^3$
(C) 82 ft$^3$
(D) 120 ft$^3$

**41.** The terminal velocity of a particle found in water is found to be 3 cm/s. The density of the solid is 3 g/cm$^3$, and the viscosity of water is 0.9 cP. The diameter of the solid is most nearly

(A) 0.016 cm
(B) 0.16 cm
(C) 1.6 cm
(D) 16 cm

**42.** The design of urban drainage facilities is based on

(A) total amount of rainfall in the area
(B) total rainfall plus any overflows from rivers
(C) total rainfall plus any overflows from lakes
(D) intensity of rainfall

**43.** A reverse-osmosis unit is used to obtain pure water from saline water. The coefficient of water permeation is 0.1 mol/cm$^2$·s·atm. The pressure differential across the membrane is 0.2 atm, and the osmotic pressure differential is 0.05 atm. The water flux across the membrane is most nearly

(A) 0.0050 mol/cm$^2$·s
(B) 0.015 mol/cm$^2$·s
(C) 0.050 mol/cm$^2$·s
(D) 0.15 mol/cm$^2$·s

**44.** The terminal drift velocity of particles to be collected in an electrostatic precipitator is 7 m/min. The flow rate of the gas stream to be treated is 60 m$^3$/s. To remove 98% of the particles, the electrostatic precipitator should have an area of most nearly

(A) 33 m$^2$
(B) 910 m$^2$
(C) 1800 m$^2$
(D) 2000 m$^2$

**45.** The effective number of turns for a high-throughput cyclone is most nearly

(A) 2.0
(B) 3.4
(C) 5.5
(D) 6.0

**46.** The criteria air pollutants for which there are National Ambient Air Quality Standards are

(A) particulate matter less than 10 $\mu$m in diameter, sulfur dioxide, nitrogen oxides, carbon monoxide, ozone, and lead
(B) sulfur dioxide, carbon monoxide, mercury, and lead
(C) volatile organic compounds, particulate matter less than 10 $\mu$m in diameter, sulfur dioxide, nitrogen oxides, carbon monoxide, and ozone
(D) sulfur dioxide, nitrogen oxides, carbon monoxide, ozone, and lead

**47.** The combustion efficiency of a particular incinerator must not drop below 99.5%. If the concentration (dry basis) of carbon dioxide leaving the incinerator is 110 000 ppm (volume), the concentration of carbon monoxide must be less than

(A) 50 ppm
(B) 300 ppm
(C) 550 ppm
(D) 700 ppm

**48.** A landfill has a 0.001 g/cm$^3$ methane concentration at the surface of its cover and a negligible methane concentration at the bottom of its cover. The depth of the landfill cover is 2 cm. The total porosity is 0.10. The flux of methane is most likely to be

(A) $5 \times 10^{-6}$ g/cm$^2$

(B) 0.1 g/cm$^2$

(C) 1 g/cm$^2$

(D) 2 g/cm$^2$

**49.** The Safe Drinking Water Act includes enforceable standards for

(A) maximum contaminant levels

(B) maximum contaminant levels and odor

(C) maximum contaminant levels and hardness and corrosivity limits

(D) maximum contaminant levels and color

**50.** Remediation of hazardous waste sites is regulated under the

(A) Comprehensive Environmental Response, Compensation, and Liability Act

(B) Emergency Planning and Community Right-to-Know Act

(C) Clean Air Act

(D) Resource Conservation and Recovery Act

**51.** An open channel has a width of 5 m and drops at the rate of 1 m/100 m. Assume a roughness coefficient, $n$, of 0.015. If the depth of the water is 3 m, the flow rate in the channel is most nearly

(A) 100 m$^3$/s

(B) 110 m$^3$/s

(C) 120 m$^3$/s

(D) 220 m$^3$/s

**52.** Excess nitrogen in composting is a problem because

(A) if there is too much nitrogen, the microorganisms cannot consume it all and it causes odor problems.

(B) the microorganisms need carbon, not nitrogen, and excess nitrogen causes the microorganisms to die off.

(C) excess nitrogen causes the composting material to cool and slows down composting.

(D) the resulting compost is too high in nitrogen to be useful.

**53.** An impeller mixer (turbulent flow) is used to rapidly incorporate a mixture with a density, $\rho_f$, of 1.20 g/mL. The mixer is of a turbine variety with six curved blades. The impeller diameter is 1 m, and the mixer rotates at 600 rpm. The mixer's power consumption is most nearly

(A) $1.0 \times 10^6$ kg·m/s

(B) $1.1 \times 10^6$ kg·m/s

(C) $5.8 \times 10^6$ kg·m/s

(D) $6.0 \times 10^6$ kg·m/s

**54.** In 2010, the population of a city was 1 million. In 2000, the population was 0.85 million. Assuming nonlinear growth, the population in 2020 is expected to be most nearly

(A) $1.0 \times 10^6$

(B) $1.2 \times 10^6$

(C) $1.4 \times 10^6$

(D) $1.6 \times 10^6$

**55.** The BOD exerted in 5 d is 100 mg/L, and the reaction rate constant is 0.1/d. The value of the ultimate BOD is most nearly

(A) 100 mg/L

(B) 200 mg/L

(C) 250 mg/L

(D) 280 mg/L

**56.** A packed tower 15 m high has a 2.5 m height of transfer unit. The number of transfer units is most nearly

(A) 2.5

(B) 6.0

(C) 15

(D) 37

**57.** If a confined aquifer at sea level has a hydraulic head of 2 m, the absolute pressure at the top of the water in the aquifer is most nearly

(A) 0.80 atm

(B) 1.0 atm

(C) 1.1 atm

(D) 1.2 atm

**58.** Potassium is an alkaline earth metal. Hazardous waste containing a potassium waste stream can safely be combined with

(A) inorganic fluorides
(B) alcohols
(C) non-oxidizing mineral acids
(D) halogenated organics

**59.** Two tanks are located in series in a wastewater treatment plant. The water level in the upstream tank is 1 m higher than that in the downstream tank. A rounded orifice with a 1 cm diameter lets water flow between the tanks. The water flow rate is most nearly

(A) 150 cm³/s
(B) 250 cm³/s
(C) 340 cm³/s
(D) 440 cm³/s

**60.** An isotherm test gives the following results.

| aqueous-phase concentration ($\mu$g/kg) | solid-phase concentration ($\mu$g/kg) |
|---|---|
| 0.05 | 0.060 |
| 0.10 | 0.120 |
| 0.15 | 0.175 |
| 0.20 | 0.240 |
| 0.30 | 0.360 |

Assume that the fraction of organic carbon in the soil is 0.15. The organic carbon partition coefficient is most nearly

(A) 0.12 kg water/kg organic carbon
(B) 0.60 kg water/kg organic carbon
(C) 4.0 kg water/kg organic carbon
(D) 8.0 kg water/kg organic carbon

**61.** A closed thermodynamic system has an initial mass of 2 kg and a total specific internal energy of 10 J/kg. The system does work amounting to 50 J. 100 J of heat is added to the system. Assume no change in kinetic energy or potential energy. The final specific internal energy of the system is most nearly

(A) 20 J/kg
(B) 35 J/kg
(C) 50 J/kg
(D) 70 J/kg

**62.** A 5 kg sample of nitrogen at a pressure of 1 atm and a temperature of 25°C needs to be compressed adiabatically to a pressure of 3 atm. The ratio of specific heats is 1.67. The reversible work done for this purpose is most nearly

(A) 350 kJ
(B) 700 kJ
(C) 1100 kJ
(D) 2000 kJ

**63.** A gas stream exits a reactor at a pressure of 0.8 atm and a temperature of 32°C. The gas stream has a mass percentage of 64% $CH_4$, 30% $C_2H_6$, and 6% $N_2$. The mass flow rate is 5 g/min. The volumetric flow rate is most nearly

(A) 0.002 m³/min
(B) 0.004 m³/min
(C) 0.006 m³/min
(D) 0.008 m³/min

**64.** The process volume relationship for a compression process in a given piston-cylinder arrangement is represented by the empirical equation $pV^{1.2} = 5.5$ kJ of heat is emitted to the surroundings during the process. The initial pressure is 1.2 atm, and the final pressure is 1.8 atm. There is no change in the kinetic energy or potential energy in the system. The change in the internal energy of the system is most nearly

(A) 500 kJ
(B) 4000 kJ
(C) 7000 kJ
(D) 10 000 kJ

**65.** Most nearly, what is the volume of the solid shown?

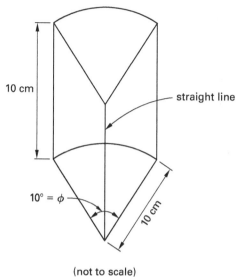

(not to scale)

(A) 50 cm³

(B) 65 cm³

(C) 87 cm³

(D) 100 cm³

**66.** Which is equal to the equation shown?

$$\frac{d}{dy}\left(\sqrt{y^5} + 2\sqrt{y}\right)$$

(A) $5y^{(4:5)} + 1$

(B) $\dfrac{1}{\sqrt{y^5}} + 2\sqrt{y}$

(C) 1

(D) $\dfrac{1}{\sqrt{y}}(2.5y^2 + 1)$

**67.** Most nearly, what is the value of the equation shown?

$$\int_4^8 \frac{2x\,dx}{x^2 + 1}$$

(A) 1.3

(B) 2

(C) 4

(D) 8

**68.** An equation is shown.

$$F(x) = 8x^4 - 64x^3 + 176x^2 - 192x + 41 = 0$$

The value of $x$ at which the maxima of the equation occurs is most nearly

(A) 0

(B) 1

(C) 2

(D) 3

**69.** In one year, a steel manufacturer purchases 4 metric tons of nickel. The facility produces 30 metric tons of stainless steel that is 8% nickel by mass, and 3 metric tons of scrap that is 8% nickel by mass. The scrap is sent to recovery. The only other significant nickel-bearing stream is nickel in waste sent to landfill. At the beginning of the year, the inventory of nickel is 0.5 metric tons; at the end of the year, it is 1 metric ton. The amount of nickel landfilled by the facility during the year is most nearly

(A) 0.86 metric ton

(B) 1.9 metric tons

(C) 2.8 metric tons

(D) 30 metric tons

**70.** An equation is shown.

$$x^3 - 2.5x - 5 = 0$$

Using the iterative Newton's method for root extraction, the value of $x$ is most nearly

(A) 2.0

(B) 2.3

(C) 2.9

(D) 3.0

**71.** An indoor facility is tested for carbon monoxide over the course of nine hours. The data from the test, expressed as a percentage of the acceptable limit for carbon monoxide in an indoor facility, is shown.

| time | concentration of CO (% of acceptable limit) |
|---|---|
| 8 am | 60 |
| 9 am | 61 |
| 10 am | 68 |
| 11 am | 68 |
| 12 pm | 70 |
| 1 pm | 70 |
| 2 pm | 70 |
| 3 pm | 71 |
| 4 pm | 72 |
| 5 pm | 70 |

The median concentration of carbon monoxide in the facility, expressed as a percentage of the acceptable limit, is most nearly

(A) 68%

(B) 69%

(C) 70%

(D) 71%

**72.** A point half a mile from a power plant is tested for the concentration of $SO_2$. The data, expressed as a percentage of the acceptable limit of $SO_2$, is shown.

| time | concentration (% of acceptable limit) |
|---|---|
| 8 am | 64 |
| 9 am | 68 |
| 10 am | 70 |
| 11 am | 70 |
| 12 pm | 72 |
| 1 pm | 82 |
| 2 pm | 70 |
| 3 pm | 70 |
| 4 pm | 68 |

The mean concentration of $SO_2$, as a percentage of the acceptable limit, is most nearly

(A) 68%

(B) 69%

(C) 70%

(D) 71%

**73.** Carbon monoxide levels are being monitored on a garage that has experienced some ventilation problems. The carbon monoxide levels at the peak time of 10 am are recorded for one week. The data, expressed as a percentage of the maximum safe level of carbon monoxide, are shown.

| day | CO (% of maximum safe level) |
|---|---|
| Sunday | 1 |
| Monday | 4 |
| Tuesday | 5 |
| Wednesday | 4 |
| Thursday | 4 |
| Friday | 3 |
| Saturday | 2 |

The value of the mode of this data is most nearly

(A) 2%

(B) 3%

(C) 4%

(D) 5%

**74.** During a theoretical study, thermal energy is drawn from a large heat reservoir by an ideal engine. The temperature in the reservoir is 120°C. The temperature of the sink from the engine is 25°C. For 1000 kJ of energy to be produced by the engine, the minimum amount of heat that needs to be drawn from the heat reservoir is most nearly

(A) 1000 kJ

(B) 2100 kJ

(C) 3300 kJ

(D) 4100 kJ

**75.** A hazardous waste containing sodium diethyldithiocarbamate may safely be combined with

(A) aliphatic amines

(B) ethanol

(C) aldehydes

(D) organic acids

**76.** The solubility of barium sulfate at close to room temperature is $1.012 \times 10^{-5}$ molar concentration in water. The ionic equilibrium concentration of barium sulfate is most nearly

(A) $10^{-10}$ g·mol/L

(B) $1.02 \times 10^{-10}$ g·mol/L

(C) $1.04 \times 10^{-10}$ g·mol/L

(D) $1.06 \times 10^{-10}$ g·mol/L

**77.** An investor receives a proposal for a factory. The proposal involves an initial investment of $50 million, annual expenses of $5 million for 10 yr, and sales of the product for 10 yr at the rate of $15,000,000 per year. The interest rate is 6%. The net future value of the project at the end of 10 years is most nearly

(A) $6,900,000

(B) $10,000,000

(C) $42,000,000

(D) $100,000,000

**78.** A pitot tube is placed in a flowing river at a depth of 0.8 m. The head in the pitot tube at the stagnation pressure is 1.5 m. The velocity of water at that point is most nearly

(A) 1.5 m/s

(B) 1.8 m/s

(C) 2.2 m/s

(D) 3.7 m/s

**79.** An open-channel flow has a sudden increase in the cross-sectional flow as a result of the width increasing at a given point. There is no pump work, and the friction loss in the channel is negligible. The height of the liquid will therefore

(A) increase

(B) decrease

(C) not change

(D) either increase, decrease, or remain unchanged depending upon environmental pressure conditions

**80.** A legislative assembly committee needs to select a subcommittee consisting of two environmental experts and three lay persons from a pool of six eligible environmental experts and six eligible lay persons. The number of ways the committee can be formed is most nearly

(A) 100

(B) 200

(C) 300

(D) 400

**81.** Residential consumption of natural gas in 2016 was 120 million m³. Assume the odorizer tetrahydrothiophene ($C_4H_8S$) was present at a density of 40 mg/m³ in residential natural gas supplies throughout the U.S. in 2016, and that the odorizer was the only significant source of sulfur in the supply. If all the sulfur in the odorizer was converted to sulfur dioxide during combustion, the mass of sulfur dioxide ($SO_2$) emitted from residential combustion of natural gas in 2016 was most nearly

(A) 1700 kg

(B) 3500 kg

(C) 4800 kg

(D) 9600 kg

**82.** An environmental project is evaluated for a fiscal benefit-cost ratio based on inherent environmental benefits. The present worth of the fiscal expenses is $5,000,000 and the present worth of the benefits is $8,125,000. The benefit-cost ratio is most nearly

(A) 0.30

(B) 0.62

(C) 1.6

(D) 3.0

**83.** An environmental project has a present cost of $20 million. The benefits provided by the project are $3,000,000 in clean water, as well as annual benefits of $3,000,000 per year for 10 yr in recycled water and reduced clean up. The interest rate is 6%. The benefit-cost ratio of the project is most nearly

(A) 1.00

(B) 1.24

(C) 1.25

(D) 1.30

**84.** An initial investment of $100,000,000 will provide a final benefit of $200,000,000 in 10 yr. The rate of return for this investment is most nearly

(A) 5%

(B) 6%

(C) 7%

(D) 8%

**85.** An investment of $10,000,000 made at the end of each year for 5 yr gives a final benefit of $60,000,000. The rate of return is most nearly

(A) 2%

(B) 5%

(C) 9%

(D) 15%

**86.** The corrosion penetration rate (CPR) in pipes is often measured using coupons made out of the same material as the pipe, immersed in the material the pipe carries. It can be calculated from the equation shown.

$$\text{CPR} = \frac{\Delta W}{\rho A t}$$

$\Delta W$ is the change in coupon weight in time $t$, $\rho$ is the density of the coupon, and $A$ is the exposed area of the coupon. If a steel coupon with an exposed area of 2.65 in$^2$ loses 33 grams per year, the CPR of the associated steel pipe is most nearly

(A) $1.0 \times 10^{-6}$ mm/yr

(B) 0.06 mm/yr

(C) 1.6 mm/yr

(D) 2.5 mm/yr

**87.** Fifty-eight humans each ingest a single dose of acetaminophen equal to 142 mg/kg of body weight. The number of expected deaths (assuming no treatment) is most nearly

(A) 0

(B) 2

(C) 29

(D) 58

**88.** A population of 2 million people breathe air that contains 4.6 ppb benzene ($C_6H_6$). Benzene is classified as a human carcinogen. The midpoint of its cancer slope factor for the inhalation route is $5 \times 10^{-6}$ per $\mu g/m^3$. The carcinogenic inhalation risk is equal to the concentration in air times the cancer slope factor for the inhalation route. The number of additional cancer cases in this population due to benzene inhalation is most nearly

(A) 4

(B) 46

(C) 160

(D) 46,000

**89.** A study of a soil sample estimates that the average concentration of glyphosate in the soil is 5 ppm by mass. A seven-year-old child ingests the soil on a daily basis; the fraction ingested is 1.0. The child's chronic daily intake of glyphosate via soil ingestion is most nearly

(A) 0.00015 ng/kg·d

(B) 0.028 ng/kg·d

(C) 0.25 ng/kg·d

(D) 7.6 ng/kg·d

**90.** Simultaneous exposure to ethanol and carbon tetrachloride has synergistic toxic effects. This means that a person who is exposed to carbon tetrachloride while drinking alcohol experiences

(A) the same toxic effects they would experience if they drank alcohol while not being exposed to carbon tetrachloride, or if they were exposed to carbon tetrachloride while not drinking alcohol, whichever is greater

(B) lower toxic effects than the toxic effects they would experience if they drank alcohol but were not exposed to carbon tetrachloride, plus the toxic effects they would experience if they were exposed to carbon tetrachloride while not drinking alcohol

(C) toxic effects equal to the toxic effects they would experience if they drank alcohol but had not been exposed to carbon tetrachloride, plus the toxic effects they would have if they had been exposed to carbon tetrachloride while not drinking alcohol

(D) greater toxic effects than the toxic effects they would experience if they drank alcohol but were not exposed to carbon tetrachloride, plus the toxic effects they would experience if they were exposed to carbon tetrachloride while not drinking alcohol

**91.** Airborne lead is present at concentrations of 0.02 $\mu g/m^3$ in an urban area. The chronic daily intake of lead via the inhalation route for a three-year-old child living year-round in this urban area is most nearly

(A) 0.44 ng/kg·d

(B) 11 ng/kg·d

(C) 26 ng/kg·d

(D) 250 ng/kg·d

**92.** For a watershed near Springfield, Illinois, the intensity of rainfall in a 100-year storm is 8.84 in/hr for a 15 min duration and 6.06 in/hr for a 30 min duration. The time of concentration in this watershed is 21 min. The value used for rainfall intensity in the rational formula for a 100-year storm in this watershed is most nearly

(A) 6.1 in/hr

(B) 7.5 in/hr

(C) 7.7 in/hr

(D) 8.8 in/hr

**93.** For a given reservoir, area and active storage volume are related as shown.

$$A = \left(\frac{2.3}{\text{km}^4}\right)V^2 + 3.1 \text{ km}^2$$

$A$ is the area of the reservoir in kilometers squared, and $V$ is its active storage volume in cubic kilometers. For this reservoir, the average net loss from evaporation less rainfall during the winter is a function of area, as shown.

$$L = \left(4.3 \times 10^{-6} \frac{\text{km}}{\text{d}}\right)A$$

$L$ is the average net loss in km$^3$/d. The active storage volume of the reservoir is 9 km$^3$ during the winter. The additional inflow into the reservoir needed to make up for the average net loss over the course of the entire winter is most nearly

(A) 0.000039 km$^3$

(B) 0.00081 km$^3$

(C) 0.074 km$^3$

(D) 0.30 km$^3$

**94.** A flood control reservoir is dry, with no inflow or outflow until after a storm begins. The inflow and outflow volumes are shown in the figure.

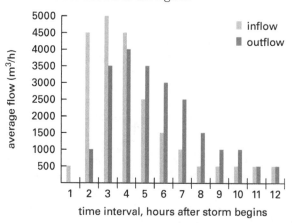

Assume no seepage or evaporation. The time interval during which the peak volume is held in the reservoir is

(A) 2 h

(B) 3 h

(C) 4 h

(D) 7 h

**95.** An empty rectangular lot with a 5% slope and a slope length of 150 ft is studied for soil erosion. The formula for the topographic factor is shown.

$$L_S = \left(\frac{\lambda}{72.6}\right)^m (65.41 \sin^2\theta + 4.56 \sin\theta + 0.065)$$

$\lambda$ is the slope length in ft, and $\theta$ is the angle of the slope. The slope length exponent, $m$, depends on the percent slope of the area under consideration, as shown in the table.

| m | slope |
|---|---|
| 0.5 | $\geq 5\%$ |
| 0.4 | 3.5% to 4.5% |
| 0.3 | 1% to 3% |
| 0.2 | $<1\%$ |

The topographic factor for the empty lot is most nearly

(A) 0.66

(B) 1.0

(C) 1.4

(D) 5.1

**96.** A channel has a maximum flow depth of 1 m and a gradient of 5%. The shear stress on the channel lining at maximum depth is given by the equation shown.

$$\tau_d = \gamma d S$$

$\gamma$ is the density of water, $d$ is the maximum flow depth, and $S$ is the channel gradient in units of m/m. The shear stress on this channel's lining at maximum depth is most nearly

(A) 3.1 kg/m$^2$

(B) 5.0 kg/m$^2$

(C) 31 kg/m$^2$

(D) 50 kg/m$^2$

**97.** An activated sludge treatment process receives a flow rate of 1.5 MGD from the primary clarifiers at a sewage treatment plant. The biochemical oxygen demand concentration in the influent is 130 mg/L. The organic loading on the activated sludge treatment plant is most nearly

(A) 23 lbf BOD/d

(B) 200 lbf BOD/d

(C) 1600 lbf BOD/d

(D) 96,000 lbf BOD/d

**98.** Conventional treatment of surface waters for drinking water will typically be adequate for removal and/or inactivation of

(A) *Giardia* and viruses

(B) *Giardia* but not viruses

(C) viruses but not *Giardia*

(D) neither *Giardia* nor viruses

**99.** A water treatment filter is operating at 20°C. The filtration rate is 0.003 m/s. The filter media is mono-sized anthracite with a depth of 0.5 m, a porosity of 0.4, and a particle size diameter of 1 mm. From the Carmen-Kozeny equation, the estimated head loss through the clean bed is most nearly

(A) 0.02 m

(B) 0.14 m

(C) 1.3 m

(D) 46 m

**100.** Two identically constructed wells are placed in separate homogeneous confined aquifers that have identical conditions. Pumping is begun at both wells at the same time. Well 1 has a pumping rate of 28 L/s. At time $t$, the difference between the drawdown at distances of 5 m and 10 m from well 1 is 0.14 m. Well 2 has a pumping rate of 46 L/s. The difference between the drawdown at distances of 5 m and 10 m from well 2 at time $t$ is most nearly

(A) 0.085 m

(B) 0.14 m

(C) 0.23 m

(D) 26 m

**101.** A confined sand aquifer has a thickness of 12 m and a specific storativity of 0.00032/m. The storativity of a confined aquifer is given by the equation shown.

$$S = S_s b$$

$S_s$ is specific storativity, and $b$ is the aquifer thickness. The storativity of the aquifer is most nearly

(A) 0.000027

(B) 0.0038

(C) 260

(D) 38,000

**102.** A well that reaches the bottom of a homogeneous unconfined aquifer has a radius of 0.15 m and a pumping rate of 32 L/s. Hydraulic head at the well is 7.5 m. Hydraulic conductivity is 0.028 m/s. The cross-sectional area of radial flow into the well is the area of a cylinder with the radius of the well and the height of the water table, as shown.

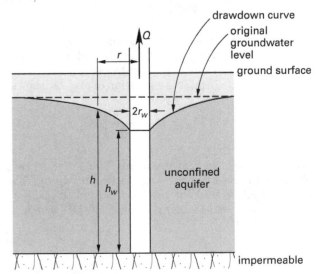

The hydraulic head 15 m away from the center of the well is most nearly

(A) 7.4 m

(B) 7.6 m

(C) 8.2 m

(D) 42 m

**103.** An air sparging system for groundwater remediation is checked to ensure the maximum air pressure does not exceed the aquifer's fracturing pressure. The fracturing pressure is the weight of the soil plus the weight of the water above the screened interval of the sparging system. The screened interval of the sparging system begins 10 m below ground. The top of the unconfined aquifer is 6 m below ground. The specific gravity of the solid phase of the soil matrix is 2.7, and the soil porosity is 0.4. A safety factor of 75% is applied (i.e., the system is designed for 75% of the fracturing pressure). Assume that the water content of the soil above the aquifer is negligible. The design sparging pressure (gage pressure) is most nearly

(A) 130 kPa

(B) 140 kPa

(C) 150 kPa

(D) 170 kPa

**104.** An unconfined aquifer with a storativity of 0.14 has an area of 400 km². Over a period of time, a net loss of 0.093 km³ stored water occurs. The drop in the water table is most nearly

(A) 0.002 m
(B) 0.14 m
(C) 0.59 m
(D) 1.7 m

**105.** Unsaturated soil at 20°C is contaminated with a pure substance. Remediation will occur via soil vapor extraction, with the substance adsorbed onto activated carbon. The adsorption behavior follows the Langmuir isotherm under the given conditions. The mass of the substance required to completely saturate a unit mass of activated charcoal is 0.33 g/g. The air flow from the well is 0.5 m³/min, and the concentration of the substance in the pumped air is constant at 10 mg/L. The constant $K$ for the Langmuir isotherm is 0.2 m³/g for an equilibrium concentration in units of g/m³ and a mass ratio in units of g/g. The desired concentration of the substance after adsorption is 2.4 g/m³. The volume of fresh activated carbon needed each day is most nearly

(A) 5.4 kg/d
(B) 36 kg/d
(C) 40 kg/d
(D) 51 kg/d

**106.** A water sample contains ionic components at the concentrations shown.

| | |
|---|---|
| $Ca^{2+}$ | 78 mg/L |
| $Mg^{2+}$ | 95 mg/L |
| $Na^+$ | 40 mg/L |

The total hardness of the water sample is most nearly

(A) 3.0 meq/L
(B) 7.8 meq/L
(C) 12.0 meq/L
(D) 14.0 meq/L

**107.** A sample of water contains ionic components at the concentrations shown. The alkalinity of the sample is 82 as $CaCO_3$, and the pH of the sample is 7.01.

| | |
|---|---|
| $Ca^{2+}$ | 90 mg/L |
| $Mg^{2+}$ | 92 mg/L |
| $Na^+$ | 18 mg/L |

The non-carbonate hardness of the sample is most nearly

(A) 4.50 meq/L
(B) 7.54 meq/L
(C) 10.4 meq/L
(D) 13.7 meq/L

**108.** A 200 mL sample of wastewater is heated to 113°C in a dish. The mass of the empty dish is 47.510 mg. The mass of the wastewater and the dish together is 47.905 mg. The dish and sample are heated to 103°C, at which point the mass of the sample and dish together is 47.760 mg. Another empty dish with a mass of 47.132 mg is filled with a 400 mL sample of the same wastewater, which is first filtered through filter paper. The dish and sample are heated to 103°C, at which point the total mass is 47.410 mg. The dish and sample are again heated, this time to 600°C, and allowed to cool slowly. The new mass of the dish and sample after cooling is 47.382 mg.

The total suspended solids in the wastewater is most nearly

(A) 0.25 mg/L
(B) 0.33 mg/L
(C) 0.63 mg/L
(D) 1.25 mg/L

**109.** A town in a large metropolitan area will be subsumed by a larger city in 10 years; when subsumed, the town will constitute 10% of the larger metropolitan area. City planners decide it may, be useful to estimate the water consumption of the town in 10 years. Current per capita consumption of water in the metropolitan area is 169 gpcd. The per-capita consumption is expected to increase by a total of 1% over the next 10 years due to additional industrialization. The population of the metro area over the past 10 years has increased by 20%, and this is expected to remain the case for the foreseeable future. The increase in the total water consumption in the town after 10 years is most nearly

(A) 20%
(B) 21%
(C) 22%
(D) 23%

**110.** What is the formation of rust on iron caused by?

(A) exposure to water vapor in the environment

(B) a reducing environment

(C) an oxidizing environment

(D) all of the above could cause rust

## SOLUTIONS

**1.** For pipes with a sharp entrance, $C$ has a value of 0.5.

$$v = \frac{Q}{A} = \frac{Q}{\frac{\pi D^2}{4}}$$

The head loss at the pipe entrance is given by the formula

$$h = \frac{Cv^2}{2g} = \frac{C\left(\frac{Q}{\frac{\pi D^2}{4}}\right)^2}{2g}$$

$$= \frac{(0.5)\left(\frac{\left(1.0\ \frac{L}{min}\right)\left(\frac{1\ m^3}{1000\ L}\right)\left(\frac{1\ min}{60\ s}\right)}{\frac{\pi(0.04\ m)^2}{4}}\right)^2}{(2)\left(9.81\ \frac{m}{s^2}\right)}$$

$$= 4.5 \times 10^{-6}\ m$$

**The answer is (C).**

**2.** For mixtures of ideal gases, the volumetric fraction and the mole fraction are the same.

$$y_{SO_2} = 0.1 \times 10^{-6}$$

Use Henry's law. The partial pressure of $SO_2$ is

*Henry's Law at Constant Temperature*

$$P_{SO_2} = P_{total}\, y_{SO_2}$$
$$= (1\ atm)(0.1 \times 10^{-6})$$
$$= 1.0 \times 10^{-7}\ atm$$

When $SO_2$ dissociates, an equal number of $HSO_3^-$ and $H_3O^+$ ions are created, so

$$[HSO_3^-][H_3O^+] = [H_3O^+]^2$$

Solve the equilibrium constant equation for the $[H_3O^+]$ concentration.

$$[H_3O^+] = \sqrt{p_{SO_2} K_{eq}}$$

$$= \sqrt{(1.0 \times 10^{-7} \text{ atm})\left(2.1 \times 10^{-2} \frac{\left(\frac{\text{mol}}{\text{L}}\right)^2}{\text{atm}}\right)}$$

$$= 4.6 \times 10^{-5} \text{ mol/L}$$

**The answer is (B).**

**3.** Alpha particles consist of helium nuclei containing two neutrons and two protons and are not capable of penetrating skin. Aluminum, lead, and concrete are even more difficult to penetrate than skin.

**The answer is (D).**

**4.** Calculate the cross-sectional areas.

$$A_1 = \pi r_1^2$$
$$= \pi \left(\frac{11 \text{ cm}}{2}\right)^2$$
$$= 95 \text{ cm}^2 \quad (9.5 \times 10^{-3} \text{ m}^2)$$
$$A_2 = \pi r_2^2$$
$$= \pi \left(\frac{5 \text{ cm}}{2}\right)^2$$
$$= 20 \text{ cm}^2 \quad (2.0 \times 10^{-3} \text{ m}^2)$$

From the diagram, the centerline of sections 1 and 2 is at the same elevation, so

$$z_1 - z_2 = 0 \text{ cm} \quad (0 \text{ m})$$

The flow rate is

$$Q = \frac{C_v A_2}{\sqrt{1 - \left(\frac{A_2}{A_1}\right)^2}} \sqrt{2g\left(\frac{p_1}{\gamma} + z_1 - \frac{p_2}{\gamma} - z_2\right)}$$

$$= \frac{C_v A_2}{\sqrt{1 - \left(\frac{A_2}{A_1}\right)^2}} \sqrt{2g\left(\frac{p_1 - p_2}{\gamma} + (z_1 - z_2)\right)}$$

$$= \frac{(0.85)(2.0 \times 10^{-3} \text{ m}^2)}{\sqrt{1 - \left(\frac{2.0 \times 10^{-3} \text{ m}^2}{9.5 \times 10^{-3} \text{ m}^2}\right)^2}}$$

$$\times \sqrt{(2)\left(9.8 \frac{\text{m}}{\text{s}^2}\right)\left(\frac{400 \text{ Pa} - 160 \text{ Pa}}{9.8 \times 10^3 \frac{\text{N}}{\text{m}^3}} + 0 \text{ m}\right)}$$

$$= 1.2 \times 10^{-3} \text{ m}^3/\text{s}$$

**The answer is (C).**

**5.** The power lost in a pipe is

$$P = Q\rho g h_f$$

For parallel pipes, the head loss, $h_f$, is the same for both pipes. Therefore,

$$\frac{P_1}{P_2} = \frac{Q_1}{Q_2}$$

The velocity in both pipes is equal, and

$$Q = vA$$

Therefore,

$$\frac{P_1}{P_2} = \frac{D_1^2}{D_2^2} = \left(\frac{4}{2}\right)^2 = 4 \quad (4.0)$$

**The answer is (C).**

**6.** Use the Streeter-Phelps equation. Solving for $t$ would be difficult, so examine the different choices to see which one results in a dissolved oxygen deficit of 0.14 mg/L. Start with $D_a = 0$ mg/L and $t = 10$ d.

**Microbial Kinetics: Stream Modeling**

$$D = \frac{k_d L_a}{k_r - k_d}[\exp(-k_d t) - \exp(-k_r t)] + D_a \exp(-k_r t)$$

$$= \left( \frac{\left(0.1 \frac{1}{d}\right)\left(7 \frac{mg}{L}\right)}{0.35 \frac{1}{d} - 0.1 \frac{1}{d}} \right)$$

$$\times \left( \exp\left[-\left(0.1 \frac{1}{d}\right)(10\ d)\right] - \exp\left[-\left(0.35 \frac{1}{d}\right)(10\ d)\right] \right)$$

$$= 0.95\ \text{mg/L}$$

Therefore, the answer is not 10 d. At $D_a = 0$ mg/L and $t = 20$ d,

$$D = \frac{k_d L_a}{k_r - k_d}[\exp(-k_d t) - \exp(-k_r t)] + D_a \exp(-k_r t)$$

$$= \left( \frac{\left(0.1 \frac{1}{d}\right)\left(7 \frac{mg}{L}\right)}{0.35 \frac{1}{d} - 0.1 \frac{1}{d}} \right)$$

$$\times \left( \exp\left[-\left(0.1 \frac{1}{d}\right)(20\ d)\right] - \exp\left[-\left(0.35 \frac{1}{d}\right)(20\ d)\right] \right)$$

$$= 0.38\ \text{mg/L}$$

Therefore, the answer is not 20 d. At $D_a = 0$ mg/L and $t = 30$ d,

$$D = \frac{k_d L_a}{k_r - k_d}[\exp(-k_d t) - \exp(-k_r t)] + D_a \exp(-k_r t)$$

$$= \left( \frac{\left(0.1 \frac{1}{d}\right)\left(7 \frac{mg}{L}\right)}{0.35 \frac{1}{d} - 0.1 \frac{1}{d}} \right)$$

$$\times \left( \exp\left[-\left(0.1 \frac{1}{d}\right)(30\ d)\right] - \exp\left[-\left(0.35 \frac{1}{d}\right)(30\ d)\right] \right)$$

$$= 0.14\ \text{mg/L}$$

This means that 30 d is the correct answer. For the sake of completeness, find $D$ at $D_a = 0$ mg/L and $t = 40$ d.

$$D = \frac{k_d L_a}{k_r - k_d}[\exp(-k_d t) - \exp(-k_r t)] + D_a \exp(-k_r t)$$

$$= \left( \frac{\left(0.1 \frac{1}{d}\right)\left(7 \frac{mg}{L}\right)}{0.35 \frac{1}{d} - 0.1 \frac{1}{d}} \right)$$

$$\times \left( \exp\left[-\left(0.1 \frac{1}{d}\right)(40\ d)\right] - \exp\left[-\left(0.35 \frac{1}{d}\right)(40\ d)\right] \right)$$

$$= 0.051\ \text{mg/L}$$

Therefore, the answer is not 40 d.

**The answer is (C).**

**7.** The relationship between the impeller diameter, $D$, and the discharge, $Q$, for a centrifugal pump at constant rotational speed (from the scaling law formula) is

**Fans, Pumps, and Compressors**

$$\frac{Q_2}{Q_1} = \left(\frac{D_2}{D_1}\right)^3$$

Solving for $D_2$ and substituting the given values yields

$$D_2 = D_1 \left(\frac{Q_2}{Q_1}\right)^{1/3}$$

$$= (1\ \text{m}) \left(\frac{20 \frac{L}{s}}{10 \frac{L}{s}}\right)^{1/3}$$

$$= 1.3\ \text{m}$$

**The answer is (C).**

**8.** The relationship between the power consumption, $P$, and impeller diameter, $D$, for a centrifugal pump at constant rotational speed (from the scaling law formula) is

**Fans, Pumps, and Compressors**

$$\frac{P_2}{P_1} = \left(\frac{D_2}{D_1}\right)^5$$

Solving for $D_2$ and substituting given values yields

$$D_2 = D_1 \left(\frac{P_2}{P_1}\right)^{1/5}$$

$$= (1 \text{ m}) \left(\frac{20 \text{ kW}}{10 \text{ kW}}\right)^{1/5}$$

$$= 1.2 \text{ m}$$

**The answer is (A).**

**9.** The terrain-adjusted stack height is determined from

$$\text{TESH} = h + \Delta h - T_r$$

Solve for $T_r$.

$$T_r = h + \Delta h - \text{TESH}$$
$$= 30 \text{ m} + 10 \text{ m} - 28 \text{ m}$$
$$= 12 \text{ m}$$

**The answer is (B).**

**10.** The ratio of minimum flow rate to the average daily flow rate of sewage can be read from a sewage flow ratio curves graph. [**Sewage Flow Ratio Curves**]

The minimum flow curve at the point where the population is 20,000 is 0.35.

**The answer is (A).**

**11.** The overall chemical equation is

$$\text{CaO} + \text{SO}_2 \rightarrow \text{CaSO}_3$$

Therefore, the molar ratio of lime to sulfur dioxide is 1:1. Find the molecular weights for lime and sulfur dioxide.

$$\text{MW}_{\text{CaO}} = (1)\left(40 \, \frac{\text{g}}{\text{mol}}\right) + (1)\left(16 \, \frac{\text{g}}{\text{mol}}\right) = 56 \text{ g/mol}$$

$$\text{MW}_{\text{SO}_2} = (1)\left(32 \, \frac{\text{g}}{\text{mol}}\right) + (2)\left(16 \, \frac{\text{g}}{\text{mol}}\right) = 64 \text{ g/mol}$$

The amount of lime required to neutralize each kilogram of sulfur dioxide is

$$\frac{56 \text{ kg CaO}}{64 \text{ kg SO}_2} = 0.88 \text{ kg CaO/kg SO}_2$$

**The answer is (A).**

**12.** The equation for combustion of benzene is

$$\text{C}_6\text{H}_6 + \text{O}_2 \rightarrow 6\text{CO}_2 + 3\text{H}_2\text{O}$$

HHV is the heat given off by the combustion reaction when there is no phase change from liquid water to steam for the water formed during combustion. Therefore, the higher heating value is

**Heats of Reaction**

$$\text{HHV} = (H_r^o) = \sum_{\text{products}} v_i (\Delta H_f^o)_i - \sum_{\text{reactants}} v_i (\Delta H_f^o)_i$$

$$= -11.7 \, \frac{\text{kcal}}{\text{mol}} + (6)\left(-94.1 \, \frac{\text{kcal}}{\text{mol}}\right)$$

$$+ (3)\left(-68.3 \, \frac{\text{kcal}}{\text{mol}}\right)$$

$$= -780 \text{ kcal/mol}$$

**The answer is (A).**

**13.** Use a hydraulic elements graph to obtain the solution. [**Hydraulic-Elements Graph for Circular Sewers**]

The ratio of depth, $d$, to diameter, $D$, must be found, as follows.

$$\frac{d}{D} = \frac{25 \text{ cm}}{40 \text{ cm}} = 0.625$$

The flow-rate curve in the chart at that depth-to-diameter ratio is 0.6, so that

$$\frac{Q}{Q_f} = 0.60$$

Solving for $Q$ and substituting in the given value for $Q_f$ yields

$$Q = 0.6 Q_f$$
$$= (0.6)\left(200 \, \frac{\text{L}}{\text{s}}\right)$$
$$= 120 \text{ L/s}$$

**The answer is (C).**

**14.** Calculate the molecular weights of formaldehyde, water, and methylene glycol.

$$MW_{HCHO} = (1)\left(12\ \frac{g}{mol}\right) + (2)\left(1\ \frac{g}{mol}\right)$$
$$+ (1)\left(16\ \frac{g}{mol}\right)$$
$$= 30\ g/mol$$

$$MW_{H_2C(OH)_2} = (1)\left(12\ \frac{g}{mol}\right) + (4)\left(1\ \frac{g}{mol}\right)$$
$$+ (2)\left(16\ \frac{g}{mol}\right)$$
$$= 48\ g/mol$$

$$MW_{H_2O} = (2)\left(1\ \frac{g}{mol}\right) + (1)\left(16\ \frac{g}{mol}\right)$$
$$= 18\ g/mol$$

Convert the concentration of formaldehyde to a molar concentration.

$$C_{HCHO} = 1\ ppb$$
$$= \left(\frac{1\ g\ HCHO}{1 \times 10^9\ g\ H_2O}\right)\left(\frac{1\ mol\ HCHO}{30\ g\ HCHO}\right)$$
$$\times \left(\frac{1\ g\ H_2O}{10^{-3}\ L\ H_2O}\right)$$
$$= 3.33 \times 10^{-8}\ mol\ HCHO/L\ H_2O$$

The equilibrium constant for this reaction is

**Equilibrium Constant of a Chemical Reaction**

$$K_{eq} = \frac{[C]^c[D]^d}{[A]^a[B]^b}$$
$$= \frac{[H_2C(OH)_2]}{[HCHO]}$$

Solve for $[H_2C(OH)_2]$.

$$[H_2C(OH)_2] = K_{eq}[HCHO]$$
$$= (2000)\left(3.33 \times 10^{-8}\ \frac{mol}{L}\right)$$
$$= 6.66 \times 10^{-5}\ mol/L$$

Convert to ppb.

$$C_{H_2C(OH)_2} = \left(6.66 \times 10^{-5}\ \frac{mol\ H_2C(OH)_2}{L\ H_2O}\right)$$
$$\times \left(48\ \frac{g\ H_2C(OH)_2}{mol\ H_2C(OH)_2}\right)\left(\frac{10^{-3}\ L\ H_2O}{1\ g\ H_2O}\right)$$
$$= 3000\ ppb\ \ (mass)$$

**The answer is (D).**

**15.** This is a mass balance problem. At steady state,

$$F_{in} = F_{out} = \frac{M}{\theta}$$

In this equation, $F$ is flux, $M$ is amount of material, and $\theta$ is residence time. The amount of material can be found from the concentration of the compound in air, the total mass of air in the atmosphere, and the molecular weights of air and of the compound, so that

$$M = \left(\frac{2\ mol\ compound}{10^6\ mol\ air}\right)(5.2 \times 10^{18}\ kg\ air)\left(10^3\ \frac{g}{kg}\right)$$
$$\times \left(\frac{1\ mol\ air}{29\ g\ air}\right)\left(16\ \frac{g\ compound}{mol\ compound}\right)$$
$$= 5.7 \times 10^{15}\ g\ compound$$

Substituting values into the equation for flux gives

$$F_{in} = \frac{5.7 \times 10^{15}\ g\ compound}{(10\ yr)\left(365\ \frac{d}{yr}\right)\left(24\ \frac{h}{d}\right)\left(3600\ \frac{s}{h}\right)}$$
$$= 20 \times 10^6\ g/s$$

**The answer is (A).**

**16.** For ideal gases, the volumetric ratio is the same as the mole ratio, since all gases have the same molar volume (i.e., 22.4 L/mol). Therefore,

$$y = (10^{-6}\ ppb)\left(10^{-9}\ \frac{1}{ppb}\right) = 10^{-15}$$

The following equation relates the mole fractions in the liquid and gas phases.

**Henry's Law at Constant Temperature**

$$P_i = Py_i = hx_i$$

$$x_i = \frac{Py_i}{h} = \frac{(1\text{ atm})(10^{-15})}{7.1\ \frac{\text{L·atm}}{\text{mol}}}$$

$$= 1.408 \times 10^{-16} \quad (1.4 \times 10^{-16})$$

**The answer is (A).**

**17.** A hydraulic elements graph can be used to obtain the solution. [**Hydraulic-Elements Graph for Circular Sewers**]

The ratio of actual flow rate, $Q$, to full-flowing flow rate, $Q_f$, can be found, as follows.

$$\frac{Q}{Q_f} = \frac{20\ \frac{\text{L}}{\text{s}}}{50\ \frac{\text{L}}{\text{s}}} = 0.4$$

The depth, $d$, to diameter, $D$, read from the chart at that flow-rate ratio is 0.48, so that

$$\frac{d}{D} = 0.48$$

Solving for $d$ and substituting the given value for $D$ yields

$$d = 0.48D$$
$$= (0.48)(50\text{ cm})$$
$$= 24\text{ cm}$$

**The answer is (C).**

**18.** Find the molecular weights of water and oxygen.

$$\text{MW}_{H_2O} = (2)\left(1\ \frac{\text{g}}{\text{mol}}\right) + (1)\left(16\ \frac{\text{g}}{\text{mol}}\right) = 18\text{ g/mol}$$

$$\text{MW}_{O_2} = (2)\left(16\ \frac{\text{g}}{\text{mol}}\right) = 32\text{ g/mol}$$

Use a basis of 1 L. The number of moles of oxygen is

$$n_{O_2} = \frac{C_{O_2}}{\text{MW}_{O_2}}$$

$$= \frac{\left(9.0\ \frac{\text{mg}}{\text{L}}\right)\left(\frac{1\text{ g}}{1000\text{ mg}}\right)}{32\ \frac{\text{g}}{\text{mol}}}$$

$$= 2.8 \times 10^{-4}\text{ mol/L}$$

The number of moles of water is

$$n_{H_2O} = \frac{C_{H_2O}}{\text{MW}_{H_2O}}$$

$$= \frac{1000\ \frac{\text{g}}{\text{L}}}{18\ \frac{\text{g}}{\text{mol}}}$$

$$= 56\text{ mol/L}$$

The mole fraction of oxygen is given by

$$x_{O_2} = \frac{n_{O_2}}{n_{O_2} + n_{H_2O}}$$

$$= \frac{2.8 \times 10^{-4}\ \frac{\text{mol}}{\text{L}}}{2.8 \times 10^{-4}\ \frac{\text{mol}}{\text{L}} + 56\ \frac{\text{mol}}{\text{L}}}$$

$$= 5.0 \times 10^{-6}$$

**The answer is (A).**

**19.** Find the molecular weight of $H_2S$.

$$\text{MW}_{H_2S} = (2)\left(1\ \frac{\text{g}}{\text{mol}}\right) + (1)\left(32\ \frac{\text{g}}{\text{mol}}\right) = 34\text{ g/mol}$$

At standard temperature and pressure, an ideal gas has 22.4 L/mol.

**Chemistry: Definitions**

$$Q_{H_2S} = C_{H_2S} Q_{\text{off-gas}}$$

$$= \left(\frac{12\text{ m}^3\text{ H}_2\text{S}}{10^6\text{ m}^3\text{ off-gas}}\right)\left(0.5\ \frac{\text{m}^3\text{ off-gas}}{\text{s}}\right)$$

$$= 6 \times 10^{-6}\text{ m}^3/\text{s}$$

Converting from volume per second to moles per second to milligrams per second is accomplished as follows

$$\left(6 \times 10^{-6}\ \frac{\text{m}^3}{\text{s}}\right)\left(\frac{1\text{ mol H}_2\text{S}}{22.4 \times 10^{-3}\text{ m}^3\text{ H}_2\text{S}}\right)$$

$$\times \left(34\ \frac{\text{g}}{\text{mol}}\right)\left(10^3\ \frac{\text{mg}}{\text{g}}\right) = 9.1\text{ mg/s}$$

**The answer is (D).**

**20.** Examination of the data shows that per-capita water consumption increased linearly, with an increase of 0.1 kg/person/d every 10 years. Therefore, estimated water consumption in 2010 will be 1.5 kg/person/d. Population growth is expected to be exponential, so the relationship between $\ln P$ and $t$ is expected to be linear. The equations that describe population are

$$P = P_0^{it}$$
$$\ln P = \ln P_0 + it$$

Make a table of values from the data.

| $P$ | $t$ | $\ln P$ |
|---|---|---|
| 100,000 | 1970 (0 yr) | 11.5 |
| 150,000 | 1980 (10 yr) | 11.9 |
| 220,000 | 1990 (20 yr) | 12.3 |

Every 10 years, $\ln P$ increases 4%. The formula is

$$\ln P = 11.5 + \left(\frac{0.4}{10 \text{ yr}}\right) t$$
$$= 11.5 + (0.04 \text{ yr}) t$$

Solving for $P$ and substituting 40 yr for $t$ yields

$$P = e^{11.5 + (0.04 \text{ yr})(40 \text{ yr})}$$
$$= 489{,}000 \text{ people}$$

Total water consumption in 2010 is

$$\text{total consumption in 2010}$$
$$= (\text{per-capita consumption}) P$$
$$= \left(1.50 \ \frac{\text{kg}}{\text{person} \cdot \text{d}}\right)(489{,}000 \text{ people})$$
$$= 7.3 \times 10^5 \text{ kg/d}$$

**The answer is (D).**

**21.** A mass balance gives

$$\text{input} - \text{output} + \text{generation} = \text{accumulation}$$

At steady-state, accumulation is 0 m³/s. Input is also 0 m³/s (air drawn into the room contains no solvent).

$$\text{output} = \text{generation} = C_{\text{room}} Q_{\text{vent}}$$

Solve for $Q_{\text{vent}}$.

$$Q_{\text{vent}} = \frac{\text{output}}{C_{\text{room}}}$$

$$= \frac{\left(0.5 \ \frac{\text{kg}}{\text{h}}\right)\left(\frac{1 \text{ h}}{3600 \text{ s}}\right)\left(10^6 \ \frac{\text{mg}}{\text{kg}}\right)}{1 \ \frac{\text{mg}}{\text{m}^3}}$$

$$= 140 \text{ m}^3/\text{s}$$

**The answer is (B).**

**22.** Activity is the number of decaying nuclei per second and is directly proportional to the number of radioactive atoms.

Half-Life

$$N = N_0 e^{-0.693 \, t/\tau}$$
$$= (500 \text{ Bq}) e^{(-0.693)(28 \text{ d}/15 \text{ d})}$$
$$= 140 \text{ Bq}$$

**The answer is (B).**

**23.** Solve the given rate equation for $t$.

$$t = -\frac{1}{\frac{0.15}{\text{d}}} \ln \frac{C(t)}{C_0}$$

$$= -\frac{1}{\frac{0.15}{\text{d}}} \ln \frac{0.25 \, C_0}{C_0}$$

$$= 9.2 \text{ d}$$

**The answer is (D).**

**24.** The measured concentration is

$$C = 1000 \text{ ppm}$$

The error in the concentration is

$$w_c = 100 \text{ ppm}$$

The square of the concentration is

$$R = C^2$$

Taking the derivative of this equation yields

$$\frac{\partial R}{\partial C} = 2C$$

From the Kline-McClintock equation,

**Measurement Uncertainty**

$$w_R = \sqrt{\left(w_1 \frac{\partial f}{\partial x_1}\right)^2 + \left(w_2 \frac{\partial f}{\partial x_2}\right)^2 + \cdots + \left(w_n \frac{\partial f}{\partial x_n}\right)^2}$$

$$= \sqrt{(w_c(2C))^2}$$
$$= w_C(2C)$$
$$= (100 \text{ ppm})((2)(1000 \text{ ppm}))$$
$$= 2 \times 10^5 \text{ ppm}^2$$

The error in the square of the concentration is 200 000 ppm².

**The answer is (D).**

**25.** The equation for Gaussian atmospheric dispersion modeling is shown.

**Atmospheric Dispersion Modeling (Gaussian)**

$$C = \frac{Q}{2\pi u \sigma_y \sigma_z} \exp\left(-\frac{1}{2}\frac{y^2}{\sigma_y^2}\right) \left( \begin{array}{c} \exp\left(-\frac{1}{2}\frac{(z-H)^2}{\sigma_z^2}\right) \\ + \exp\left(-\frac{1}{2}\frac{(z+H)^2}{\sigma_z^2}\right) \end{array} \right)$$

The standard deviations (SD) can be read from plume SD diagrams and are $\sigma_z = 120$ m and $\sigma_y = 220$ m at $x = 2000$ m and atmosphere class C. [**Vertical Standard Deviations of a Plume**] [**Horizontal Standard Deviations of a Plume**]

At the centerline at ground level, $y$ and $z$ are 0 and the equation simplifies to

$$C = \left(\frac{Q}{\pi u \sigma_y \sigma_z}\right) \exp\left(-\frac{1}{2}\right)\left(\frac{H^2}{\sigma_z^2}\right)$$

$$= \frac{\left(0.8 \frac{\text{kg}}{\text{s}}\right)\left(10^6 \frac{\text{mg}}{\text{kg}}\right)}{\pi\left(4 \frac{\text{m}}{\text{s}}\right)(120 \text{ m})(220 \text{ m})} \exp\left(-\frac{1}{2}\right)\left(\frac{(40 \text{ m})^2}{(120 \text{ m})^2}\right)$$

$$= 2 \text{ mg/m}^3$$

**The answer is (B).**

**26.** Chemical oxygen demand is the amount of oxygen required to convert a compound to carbon dioxide, ammonia, and water. The balanced chemical equation for this conversion is

$$C_4H_5NO_2 + 3.5O_2 \rightarrow 4CO_2 + NH_3 + H_2O$$

Calculate the molecular weights of $C_4H_5NO_2$ and oxygen.

$$\text{MW}_{C_4H_5NO_2} = (4)\left(12 \frac{\text{g}}{\text{mol}}\right) + (5)\left(1 \frac{\text{g}}{\text{mol}}\right)$$
$$+ (1)\left(14 \frac{\text{g}}{\text{mol}}\right) + (2)\left(16 \frac{\text{g}}{\text{mol}}\right)$$
$$= 99 \text{ g/mol}$$

$$\text{MW}_{O_2} = (2)\left(16 \frac{\text{g}}{\text{mol}}\right)$$
$$= 32 \text{ g/mol}$$

For every mol of $C_4H_5NO_2$, 3.5 mols of oxygen are required (from the balanced chemical equation). Therefore,

$$\text{COD} = \frac{\text{mass of } O_2 \text{ required for conversion}}{\text{mass of compound}}$$

$$= \frac{(3.5)\left(32 \frac{\text{g}}{\text{mol}}\right)}{(1)\left(99 \frac{\text{g}}{\text{mol}}\right)}$$

$$= 1.1 \text{ g } O_2/\text{g } C_4H_5NO_2$$

**The answer is (C).**

**27.** Tin and steel are not suitable. Tungsten is not suitable and expensive. Alloys of aluminum or chromium do not react with concentrated nitric acid.

**The answer is (C).**

**28.** This is a mass balance problem. The outlet flow equals the inlet flows, and the inlet ratios multiplied by their flow rates equal the outlet ratio multiplied by its flow rate.

Express the carbon: nitrogen mass ratios mathematically. The mass fraction of the mixed stream that is grass clippings is $x$.

$$\frac{m_{C,\text{grass}}}{m_{N,\text{grass}}} = 15$$

$$\frac{m_{C,\text{brush}}}{m_{N,\text{brush}}} = 55$$

$$\frac{m_{C,\text{overall}}}{m_{N,\text{overall}}} = \frac{m_{C,\text{brush}}}{m_{N,\text{brush}}}(1-x) + \frac{m_{C,\text{grass}}}{m_{N,\text{grass}}}x = 30$$

Solve for $x$.

$$\frac{m_{C,\text{brush}}}{m_{N,\text{brush}}} - \frac{m_{C,\text{brush}}}{m_{N,\text{brush}}}x + \frac{m_{C,\text{grass}}}{m_{N,\text{grass}}}x = \frac{m_{C,\text{overall}}}{m_{N,\text{overall}}}$$

$$-\frac{m_{C,\text{brush}}}{m_{N,\text{brush}}}x + \frac{m_{C,\text{grass}}}{m_{N,\text{grass}}}x = \frac{m_{C,\text{overall}}}{m_{N,\text{overall}}} - \frac{m_{C,\text{brush}}}{m_{N,\text{brush}}}$$

$$\frac{\dfrac{m_{C,\text{overall}}}{m_{N,\text{overall}}} - \dfrac{m_{C,\text{brush}}}{m_{N,\text{brush}}}}{-\dfrac{m_{C,\text{brush}}}{m_{N,\text{brush}}} + \dfrac{m_{C,\text{grass}}}{m_{N,\text{grass}}}} = \frac{30 - 55}{-55 + 15}$$

$$= 0.63 \quad (63\%)$$

**The answer is (C).**

**29.** Find the rate of removal using the half-life formula.

*Half-Life*

$$k = \frac{0.693}{t_{1/2}} = \frac{0.693}{3 \text{ wk}}$$
$$= 0.231/\text{wk}$$

After four weeks, the amount remaining is

$$Q = Q_0 e^{-kt}$$
$$= (100 \text{ kg})^{-\left(0.231 \frac{1}{\text{wk}}\right)(4 \text{ wk})}$$
$$= 40 \text{ kg}$$

**The answer is (D).**

**30.** Use the Streeter-Phelps stream modeling equation for determining the time that corresponds with the minimum dissolved oxygen concentration.

The initial dissolved oxygen deficit is

*Microbial Kinetics: Stream Modeling*

$$DO = DO_{\text{sat}} - D$$
$$= 9.1 \frac{\text{mg}}{\text{L}} - 8.0 \frac{\text{mg}}{\text{L}}$$
$$= 1.1 \text{ mg/L}$$

The time after discharge at which the water will reach its minimum dissolved oxygen concentration is then

$$t_c = \frac{1}{k_a - k_d} \ln\left(\left(\frac{k_a}{k_d}\right)\left(1 - D_0\left(\frac{k_a - k_d}{k_d S_0}\right)\right)\right)$$

$$= \frac{1}{\dfrac{4}{d} - \dfrac{0.1}{d}} \ln\left(\left(\dfrac{\dfrac{4}{d}}{\dfrac{0.1}{d}}\right) \times \left(\dfrac{1 - \left(1.1 \dfrac{\text{mg}}{\text{L}}\right)}{\left(\dfrac{0.1}{d}\right)\left(200 \dfrac{\text{mg}}{\text{L}}\right)}\left(\dfrac{4}{d} - \dfrac{0.1}{d}\right)\right)\right)$$

$$= 0.88 \text{ d} \quad (21 \text{ h})$$

**The answer is (C).**

**31.** Mass balance for an incompressible fluid gives

$$\text{rate of volume increase} = \text{inflow} - \text{outflow}$$

In the case of this lake, the rate of volume increase is

$$\frac{dV}{dt} = Q_{\text{in}} = A\frac{dh}{dt} = 50 \text{ m}^3/\text{s}$$

Integrate to get

$$\int_{10 \text{ m}}^{11 \text{ m}} \left(3 \times 10^6 \text{ m}^2 + (2 \times 10^4 \text{ m})h\right) dh$$
$$= \int_0^{t_{4\text{ m}}} \left(50 \frac{\text{m}^3}{\text{s}}\right) dt$$

$$(3 \times 10^6 \text{ m}^2)h + \left(\frac{1}{2}\right)(2 \times 10^4 \text{ m})h^2 \Big|_{10 \text{ m}}^{11 \text{ m}}$$
$$= \left(50 \frac{\text{m}^3}{\text{s}}\right)t \Big|_0^{t_{4\text{ m}}}$$

$$= \left(30 \times 10^6 \text{ m}^3 + \left(\frac{1}{2}\right)(200 \times 10^4 \text{ m}^3)\right)$$
$$\quad - \left(33 \times 10^6 \text{ m}^3 + \left(\frac{1}{2}\right)(242 \times 10^4 \text{ m}^3)\right)$$
$$= -\left(50 \frac{\text{m}^3}{\text{s}}\right) t_{4\text{ m}}$$

Solve for $t_{4\,m}$.

$$t_{4\,m} = \left(\frac{1}{50\,\frac{m^3}{s}}\right)(3.2 \times 10^6\,m^3)\left(\frac{1\,h}{3600\,s}\right)\left(\frac{1\,d}{24\,h}\right)$$
$$= 0.74\,d$$

**The answer is (B).**

**32.** The TCLP is an extraction procedure that was designed to mimic the potential for landfill leachate to extract certain compounds from their original matrix.

**The answer is (C).**

**33.** The autoionization constant of water is given by

$$K_w = [H_3O^+][OH^-]$$

The acid ionization constant for acetic acid is

$$K_a = \frac{[H_3O^+][C_2H_3O_2^-]}{[HC_2H_3O_2]}$$

The base ionization constant of acetic acid is

$$K_b = \frac{[HC_2H_3O_2][OH^-]}{[C_2H_3O_2^-]}$$

Substitution yields

$$K_w = \frac{K_a[HC_2H_3O_2]K_b[C_2H_3O_2^-]}{[C_2H_3O_2^-][HC_2H_3O_2]}$$
$$= K_a K_b$$

Solve for $K_b$.

$$K_b = \frac{K_w}{K_a}$$
$$= \frac{1 \times 10^{-14}}{1.77 \times 10^{-5}}$$
$$= 5.6 \times 10^{-10}$$

**The answer is (C).**

**34.** The compression of baling reduces the volume of solid waste.

**The answer is (D).**

**35.** The distance from the nearest point-of-use water source has no influence on the potential for landfill leachate to contaminate groundwater. Other factors being equal, the potential for leachate to contaminate groundwater decreases with increasing distance to the nearest aquifer. Higher precipitation rates increase the likelihood of groundwater contamination. Increasing the permeability of soil at the landfill results in an increased potential for groundwater contamination by landfill leachate.

**The answer is (B).**

**36.** The volumetric flux, $J_w$, can be found using the formula for ultrafiltration.

**Ultrafiltration**

$$J_w = \frac{\varepsilon r^2 \int \Delta p}{8\mu\delta}$$

$$= \frac{(0.50)(2 \times 10^{-6}\,m)^2(10\,000\,Pa)\left(\frac{1\,\frac{kg}{m\cdot s^2}}{1\,Pa}\right)}{(8)(0.5\,cP)\left(10^{-3}\,\frac{\frac{kg}{m\cdot s}}{cP}\right)(100 \times 10^{-6}\,m)}$$

$$= 0.05\,m/s$$

**The answer is (D).**

**37.** The Reynolds number is the ratio of inertial forces to friction forces; it is used in all flow scale-up problems. The Weber number is the ratio of inertial forces to surface forces; it is used in those cases where surface tension has an important bearing on the flow. This is not the case here. The Froude number is the ratio of inertial forces to gravitational forces. Gravitational forces are important here. Therefore, both the Reynolds number and Froude number are used in the given problem.

**The answer is (C).**

**38.** Use Darcy's law. The change in hydraulic head (which decreases in the direction of flow) over the length of the soil sample is

$$\frac{dh}{dx} = \frac{0.1\,ft - 3.5\,ft}{0.5\,ft} = -6.8\,ft/ft$$

The hydraulic conductivity is calculated.

### Darcy's Law

$$Q = -KA(dh/dx)$$

$$K = -\frac{Q}{A\dfrac{dh}{dx}}$$

$$= -\frac{1\,\dfrac{\text{ft}^3}{\text{day}}}{(0.07\,\text{ft}^2)\left(-6.8\,\dfrac{\text{ft}}{\text{ft}}\right)\left(24\,\dfrac{\text{hr}}{\text{day}}\right)\left(3600\,\dfrac{\text{sec}}{\text{hr}}\right)}$$

$$= 2.4 \times 10^{-5}\,\text{ft/sec}$$

**The answer is (A).**

**39.** Use the Hazen-Williams equation to calculate the velocity. The Hazen-Williams roughness coefficient for concrete is 130.

### Open-Channel Flow and/or Pipe Flow of Water

$$v = k_1 C R_H^{0.63} S^{0.54}$$
$$= (0.849)(130)(1\,\text{m})^{0.63}(0.8)^{0.54}$$
$$= 97.8\,\text{m/s} \quad (98\,\text{m/s})$$

**The answer is (C).**

**40.** Use the formula for standard rate.

### Anaerobic Digestion

$$\text{Reactor volume} = \frac{V_1 + V_2}{2} t_r + V_2 t_s$$

$$= \frac{\left(\dfrac{350\,\dfrac{\text{ft}^3}{\text{day}} + 1\,\dfrac{\text{ft}^3}{\text{day}}}{2}\right)(8\,\text{hr}) + \left(1\,\dfrac{\text{ft}^3}{\text{day}}\right)(12\,\text{hr})}{24\,\dfrac{\text{hr}}{\text{day}}}$$

$$= 59\,\text{ft}^3$$

**The answer is (B).**

**41.** Stokes' law can be used for small particles.

### Settling Equations: Stokes' Law

$$v_t = \frac{g(\rho_p - \rho_f)d^2}{18\mu}$$

From the problem statement, the density of the particle is 3 g/cm³ and viscosity is

$$\mu = 0.9\,\text{cP} = 0.9 \times 10^{-2}\,\dfrac{\text{g}}{\text{cm}\cdot\text{s}}$$

Substituting values and solving for $d$ yields

$$d = \sqrt{\frac{18\mu v_t}{g(\rho_p - \rho_f)}}$$

$$= \sqrt{\frac{(18)\left(0.9 \times 10^{-2}\,\dfrac{\text{g}}{\text{cm}\cdot\text{s}}\right)\left(3\,\dfrac{\text{cm}}{\text{s}}\right)}{\left(980\,\dfrac{\text{cm}}{\text{s}^2}\right)\left(3\,\dfrac{\text{g}}{\text{cm}^3} - 1\,\dfrac{\text{g}}{\text{cm}^3}\right)}}$$

$$= \sqrt{2.47 \times 10^{-4}\,\text{cm}^2}$$
$$= 0.016\,\text{cm}$$

**The answer is (A).**

**42.** The design of urban drainage facilities is based upon the depth of water per unit time collected during rainfall. This is known as rainfall intensity.

**The answer is (D).**

**43.** Use the formula for water flux in reverse osmosis.

### Reverse Osmosis

$$J_w = W_p(\Delta p - \Delta \pi)$$
$$= \left(0.1\,\dfrac{\text{mol}}{\text{cm}^2\cdot\text{s}\cdot\text{atm}}\right)(0.2\,\text{atm} - 0.05\,\text{atm})$$
$$= 0.015\,\text{mol/cm}^2\cdot\text{s}$$

**The answer is (B).**

**44.** The Deutsch-Anderson equation is

### Electrostatic Precipitator Efficiency

$$\eta = 1 - e^{(-WA/Q)}$$

Solve for $A$.

$$\ln(1-\eta) = -\frac{WA}{Q}$$

$$A = -\frac{Q}{W}\ln(1-\eta)$$

$$= -\frac{\left(60\,\dfrac{\text{m}^3}{\text{s}}\right)\left(60\,\dfrac{\text{s}}{\text{min}}\right)}{7\,\dfrac{\text{m}}{\text{min}}}\ln(1 - 0.98)$$

$$= 2000\,\text{m}^2$$

**The answer is (D).**

**45.** Use a table with cyclone ratio of dimensions to body diameter and the equation for approximating the effective number of turns. If body diameter is $D$, then

$$H = 0.8D$$
$$L_b = 1.7D$$
$$L_c = 2.00D$$

**Cyclone Effective Number of Turns Approximation**

$$N_e = \frac{1}{H}\left(L_b + \frac{L_c}{2}\right)$$
$$= \frac{1}{0.8D}\left(1.7D + \frac{2.00D}{2}\right)$$
$$= \left(\frac{1}{0.8}\right)\left(1.7 + \frac{2.00}{2}\right)$$
$$= 3.4$$

*The answer is (B).*

**46.** The criteria air pollutants are particulate matter less than 10 μm in diameter, sulfur dioxide, nitrogen oxides, carbon monoxide, ozone, and lead. Volatile organic compounds have an effect on ozone concentrations and are sometimes regulated under the Clean Air Act, but there are no NAAQSs for volatile organic compounds.

*The answer is (A).*

**47.** Use the equation for combustion efficiency

**Incineration**

$$CE = \frac{CO_2}{CO_2 + CO} \times 100\%$$

Solve for CO.

$$CO_2 + CO = \frac{CO_2}{CE} \times 100\%$$

$$CO = \frac{CO_2}{CE} \times 100\% - CO_2$$
$$= \frac{110\,000 \text{ ppm}}{99.5\%} \times 100\% - 110\,000 \text{ ppm}$$
$$= 550 \text{ ppm} \quad [\text{volume}]$$

*The answer is (C).*

**48.** Use the formula for the gas flux. [**Gas Flux**]

The diffusion coefficient of methane is $D = 0.20$ cm/s. The depth of landfill cover is $L = 2$ cm. Total porosity is 0.10. The values of $C_{A,\text{fill}}$ and $C_{A,\text{atm}}$ are 0 g/cm$^3$ and 0.001 g/cm$^3$, respectively. The value of the flux of methane can be computed using the formula.

**Gas Flux**

$$N_A = \frac{D\eta^{4/3}(C_{A,\text{atm}} - C_{A,\text{fill}})}{L}$$

$$= \frac{\left(0.20\,\frac{\text{cm}^2}{\text{s}}\right)(0.1)^{4/3}\left(0.001\,\frac{\text{g}}{\text{cm}^3} - 0\,\frac{\text{g}}{\text{cm}^3}\right)}{2 \text{ cm}}$$

$$= 5 \times 10^{-6} \text{ g/cm}^2$$

*The answer is (A).*

**49.** There are two types of drinking water standards: primary and secondary. The primary standards specify the maximum allowable contaminant levels based on health criteria. These standards are enforceable. The secondary standards related to odor, color, hardness, and corrosivity are not enforceable.

*The answer is (A).*

**50.** Remediation of hazardous waste sites is regulated under the Comprehensive Environmental Response, Compensation, and Liability Act.

*The answer is (A).*

**51.** Use Manning's formula.

**Manning's Equation**

$$Q = (K/n)AR_H^{2/3}S^{1/2}$$
$$= \frac{1.0}{N}AR^{2/3}S^{1/2}$$

From the problem statement, the roughness coefficient is 0.015. The cross-sectional area is

$$A = WD$$
$$= (3 \text{ m})(5 \text{ m})$$
$$= 15 \text{ m}^2$$

The wetted perimeter is

$$P = D + W_1 + W_2$$
$$= 5 \text{ m} + 3 \text{ m} + 3 \text{ m}$$
$$= 11 \text{ m}$$

Therefore, the hydraulic radius is

$$R = \frac{A}{P} = \frac{15 \text{ m}^2}{11 \text{ m}} = 1.36 \text{ m}$$

The slope is

$$S = \frac{1 \text{ m}}{100 \text{ m}} = 10^{-2} \text{ m/m}$$

Therefore, the flow rate is

$$Q = \left(\frac{1.0}{0.015}\right)(15 \text{ m}^2)(1.36 \text{ m})^{2/3}\left(10^{-2}\ \frac{\text{m}}{\text{m}}\right)^{1/2}$$
$$= 120 \text{ m}^3/\text{s}$$

**The answer is (C).**

**52.** Composting organisms need a number of nutrients to perform optimally, including both carbon and nitrogen. Nitrogen in compost is desirable. Excess nitrogen would, if anything, create more heat during the composting process, not cooling. When nitrogen is present in excess, the microorganisms cannot metabolize all of it, and it is released in the form of gases with an unpleasant odor.

**The answer is (A).**

**53.** The mixer speed is

$$n = \frac{600\ \dfrac{\text{rev}}{\text{min}}}{60\ \dfrac{\text{s}}{\text{min}}}$$
$$= 10 \text{ rev/s}$$

The diameter, $D_i$, is 1 m, and the density, $\rho_f$, is 1.20 g/mL ($1.20 \times 10^3$ kg/m$^3$). The impeller constant, $K_T$, is 4.80. [**Values of the Impeller Constant $K_T$**]

Therefore, power is

$$P = K_T n^3 D_i^5 \rho_f$$
$$= (4.80)\left(10\ \frac{\text{rev}}{\text{s}}\right)^3 (1 \text{ m}^3)^5 \left(1.20 \times 10^3\ \frac{\text{kg}}{\text{m}^3}\right)$$
$$= 5.8 \times 10^6 \text{ kg·m/s}$$

**The answer is (C).**

**54.** Since the growth is nonlinear, the exponential growth curve can be used. At time $t = 0$ yr (2000), the population, $P_0$, is 0.85 million; at time $t = 10$ yr (2010), the population, $P_{10}$, is 1.00 million. Therefore,

$$\ln \frac{P_{10}}{P_0} = k\Delta t$$

Solve for $k$.

$$k = \frac{\ln \dfrac{P_{10}}{P_0}}{\Delta t}$$
$$= \frac{\ln \dfrac{1.00 \text{ million}}{0.85 \text{ million}}}{10 \text{ yr}}$$
$$= 1.625 \times 10^{-2}/\text{yr}$$

Therefore, the population in 2020 ($t = 20$ yr) is

$$P_{20} = P_0 e^{kt}$$
$$= (0.85 \times 10^6) e^{\left(\frac{1.625 \times 10^{-2}}{\text{yr}}\right)(20 \text{ yr})}$$
$$= 1.2 \times 10^6$$

**The answer is (B).**

**55.** The relationship used is

$$y_t = L(1 - e^{-kt})$$

Solve for $L$, the ultimate BOD.

$$L = \frac{y_t}{1 - e^{-kt}}$$

The BOD exerted in 5 d, $y_5$, is 100 mg/L.

$$L = \frac{100\ \dfrac{\text{mg}}{\text{L}}}{1 - e^{-(0.1/\text{d})(5 \text{ d})}}$$
$$= 250 \text{ mg/L}$$

**The answer is (C).**

**56.** The height of the tower is

**Air Stripping: Stripper Packing Height = Z**
$$Z = \text{HTU} \times \text{NTU}$$

The height of the transfer units, HTU, is 2.5 m. Solving for NTU and substituting values yields

$$\text{NTU} = \frac{Z}{\text{HTU}}$$
$$= \frac{15 \text{ m}}{2.5 \text{ m}}$$
$$= 6.0$$

**The answer is (B).**

**57.** The gage pressure of the water at the top of the aquifer can be found from the hydraulic head.

$$p_w = \rho g h$$
$$= \left(1000 \ \frac{\text{kg}}{\text{m}^3}\right)\left(9.8 \ \frac{\text{m}}{\text{s}^2}\right)(2 \text{ m})\left(9.869 \times 10^{-6} \ \frac{\text{atm}}{\text{Pa}}\right)$$
$$= 0.19 \text{ atm}$$

The absolute pressure at the top of the water in the aquifer is the atmospheric pressure plus the pressure from 2 m of water.

$$p_{\text{total}} = p_{\text{atm}} + p_w$$
$$= 1.0 \text{ atm} + 0.19 \text{ atm}$$
$$= 1.2 \text{ atm}$$

**The answer is (D).**

**58.** From a hazardous waste compatibility chart, mixing alkaline earth metals with alcohols or non-oxidizing minerals might generate flammable gas, heat, and fire. [**Hazardous Waste Compatibility Chart**]

Mixing alkaline earth metals with halogenated organics would generate heat and explosion. There are no consequences from mixing alkaline earth metals with inorganic fluorides.

**The answer is (A).**

**59.** The coefficient of the orifice is 0.98 because it is a rounded orifice. [**Orifices**]

The area of the orifice is

$$A = \pi \frac{d^2}{4} = \pi \left(\frac{(1 \text{ cm})^2}{4}\right) = 0.785 \text{ cm}^2$$

The difference in the height of the water level in the two tanks is

$$h_1 - h_2 = 1 \text{ m} \quad (100 \text{ cm})$$

For a submerged orifice, the volumetric flow rate is given by

$$Q = CA\sqrt{2g(h_1 - h_2)}$$
$$= (0.98)(0.785 \text{ cm}^2)\sqrt{(2)\left(980 \ \frac{\text{cm}}{\text{s}^2}\right)(100 \text{ cm})}$$
$$= 340 \text{ cm}^3/\text{s}$$

**The answer is (C).**

**60.** Analysis of the data shows that the soil-water partition coefficient is

**Soil-Water Partition Coefficient** $K_{sw} = K_p$

$$K_{sw} = X/C = K_{oc}f_{oc}$$
$$= \frac{0.060 \ \frac{\mu\text{g compound}}{\text{kg soil}}}{0.05 \ \frac{\mu\text{g compound}}{\text{kg water}}}$$
$$= 1.2 \text{ kg water/kg soil}$$

Solving for $K_{oc}$, the organic carbon partition coefficient, and substituting in the value for $f_{oc}$, the fraction of organic carbon in the soil, yields

$$K_{oc} = \frac{K_{sw}}{f_{oc}}$$
$$= \frac{1.2 \ \frac{\text{kg water}}{\text{kg soil}}}{0.15 \ \frac{\text{kg organic carbon}}{\text{kg soil}}}$$
$$= 8.0 \text{ kg water/kg organic carbon}$$

**The answer is (D).**

**61.** For a closed thermodynamic system, the equation for the difference between heat added to a system and work done by a system is

**Closed Thermodynamic System**
$$Q - W = \Delta U + \Delta KE + \Delta PE$$

$\Delta U$ is the change in internal energy, $\Delta KE$ is the change in kinetic energy, and $\Delta PE$ is the change in potential energy. There is no change in internal energy and potential energy, so $\Delta KE = \Delta PE = 0$. The change in the internal energy is

$$\Delta U = Q - W = 100 \text{ J} - 50 \text{ J} = 50 \text{ J}$$

This can also be expressed in terms of the final internal energy and the initial internal energy.

$$U_f - U_i = 50 \text{ J}$$

From the equation for specific internal energy, the initial internal energy of the system is

**State Functions (Properties)**
$$u = U/m$$
$$U_i = u_i m$$
$$= \left(10 \ \frac{\text{J}}{\text{kg}}\right)(2 \text{ kg})$$
$$= 20 \text{ J}$$

The final internal energy is

$$U_f - U_i = 50 \text{ J}$$
$$U_f = 50 \text{ J} + U_i$$
$$= 50 \text{ J} + 20 \text{ J}$$
$$= 70 \text{ J}$$

The final specific internal energy is

$$u_f = U_f/m$$
$$= \frac{70 \text{ J}}{2 \text{ kg}}$$
$$= 35 \text{ J/kg}$$

**The answer is (B).**

**62.** Because the process is adiabatic, it is also isentropic. In an isentropic process, the relationship between pressure and volume is

**Special Cases of Closed Systems**

$$PV^{\kappa} = \text{constant}$$

$$k = C_p/C_v = 1.67$$

The equation for the work done is

$$W = \int \delta W = \frac{(P_2 V_2 - P_1 V_1)}{-(1 - \kappa)}$$

Find the initial volume. From the ideal gas law, the equation for the volume is

$$V_1 = \frac{nRT_1}{P_1}$$

Find the number of moles.

$$n = \frac{5 \text{ kg}}{\text{MW N}_2} = 178.6 \text{ mol}$$

$$= \frac{(5 \text{ kg})\left(1000 \dfrac{\text{g}}{\text{kg}}\right)}{28.0134 \dfrac{\text{g}}{\text{mol}}}$$

$$= 178.6 \text{ mol}$$

The initial absolute temperature is

$$T_1 = 25°\text{C} + 273° = 298\text{K}$$

The initial volume is

$$V_1 = \frac{(178.6 \text{ mol})\left(8.314 \dfrac{\text{J}}{\text{mol·K}}\right)(298\text{K})}{(1 \text{ atm})\left(101.3 \dfrac{\text{kPa}}{\text{atm}}\right)\left(1000 \dfrac{\text{Pa}}{\text{kPa}}\right)} = 4.36 \text{ m}^3$$

$P_2 V_2$ is related to the product of the initial pressure and initial volume as shown.

$$P_1 V_1^{1.67} = P_2 V_2^{1.67}$$

The final volume is

$$V_2 = \left(\frac{P_1 V_1^{1.67}}{P_2}\right)^{1/1.67} = \left(\frac{(1 \text{ atm})(4.36 \text{ m}^3)^{1.67}}{3 \text{ atm}}\right)^{1/1.67} = 2.22 \text{ m}^3$$

The work done is

$$W = \left(\frac{(3 \text{ atm})(2.22 \text{ m}^3) - (1 \text{ atm})(4.36 \text{ m}^3)}{-(1 - 1.67)}\right)\left(101.3 \dfrac{\text{kPa}}{\text{atm}}\right)$$
$$= 347 \text{ kJ} \quad (350 \text{ kJ})$$

**The answer is (A).**

**63.** The mass of $CH_4$ in 5 g of the gas stream is

$$m_{CH_4} = (5 \text{ g})(0.64) = 3.2 \text{ g}$$

The mass of $C_2H_6$ in 5 g of the gas stream is

$$m_{C_2H_6} = (5 \text{ g})(0.30) = 1.5 \text{ g}$$

The mass of $N_2$ in 5 g of the gas stream is

$$m_{N_2} = (5 \text{ g})(0.06) = 0.30 \text{ g}$$

The molecular weight of $CH_4$ is

$$MW_{CH_4} = (1)\left(12 \dfrac{\text{g}}{\text{mol}}\right) + (4)\left(1 \dfrac{\text{g}}{\text{mol}}\right) = 16 \text{ g/mol}$$

The molecular weight of $C_2H_6$ is

$$MW_{C_2H_6} = (2)\left(12 \dfrac{\text{g}}{\text{mol}}\right) + (6)\left(1 \dfrac{\text{g}}{\text{mol}}\right) = 30 \text{ g/mol}$$

The molecular weight of $N_2$ is

$$MW_{N_2} = (2)\left(14 \dfrac{\text{g}}{\text{mol}}\right) = 28 \text{ g/mol}$$

The number of moles of each component in 5 g of the gas stream is

$$N_{CH_4} = \frac{m_{CH_4}}{MW_{CH_4}} = \frac{3.2 \text{ g}}{16 \frac{\text{g}}{\text{mol}}} = 0.2 \text{ g·mol}$$

$$N_{C_2H_6} = \frac{m_{C_2H_6}}{MW_{C_2H_6}} = \frac{1.5 \text{ g}}{30 \frac{\text{g}}{\text{mol}}} = 0.05 \text{ g·mol}$$

$$N_{N_2} = \frac{m_{N_2}}{MW_{N_2}} = \frac{0.30 \text{ g}}{28 \frac{\text{g}}{\text{mol}}} = 0.0107 \text{ g·mol}$$

The flow rate of the stream is 5 g/min, so the flow rate of each chemical in the gas stream is equal to the number of moles in 5 g of the gas stream per minute. The molar flow rate of the stream is

$$0.2 \frac{\text{g·mol}}{\text{min}} + 0.05 \frac{\text{g·mol}}{\text{min}} + 0.0107 \frac{\text{g·mol}}{\text{min}} = 0.2607 \text{ g·mol/min}$$

The absolute temperature is

$$T = 32°C + 273° = 305 \text{K}$$

From the ideal gas law, the volumetric flow rate is

$$pV = nRT$$
$$V = \frac{nRT}{p}$$
$$= \frac{\left(0.2607 \frac{\text{g·mol}}{\text{min}}\right)\left(8.314 \frac{\text{J}}{\text{mol·K}}\right)(305\text{K})}{(0.8 \text{ atm})\left(101.3 \frac{\text{kPa}}{\text{atm}}\right)\left(1000 \frac{\text{Pa}}{\text{kPa}}\right)}$$
$$= 0.00815 \text{ m}^3/\text{min} \quad (0.008 \text{ m}^3/\text{min})$$

**The answer is (D).**

**64.** From the process volume relationship, the equation for the pressure is

$$pV^{1.2} = 5$$
$$p = \frac{5}{V^{1.2}}$$

The equation for the work done by the system is

$$W = M \int_{V_i}^{V_f} p \, dV$$
$$= M \int_{V_i}^{V_f} \frac{5}{V^{1.2}} dV$$

The initial volume is

$$V_i = \left(\frac{5}{p_i}\right)^{1/1.2} = \left(\frac{5}{1.2 \text{ atm}}\right)^{1/1.2} = 3.27 \text{ m}^3$$

The final volume is

$$V_i = \left(\frac{5}{p_f}\right)^{1/1.2} = \left(\frac{5}{1.8 \text{ atm}}\right)^{1/1.2} = 2.33 \text{ m}^3$$

The work done is

$$W = M \int_{V_i}^{V_f} \frac{5}{V^{1.2}} dV = (5)\left(\frac{V^{-1.2+1}}{(-1.2+1)}\right)_{V_i}^{V_f}$$
$$= (5)(-5 \text{ kJ})\left(\frac{1}{(2.33 \text{ m}^3)^{0.2}} - \frac{1}{(3.27 \text{ m}^3)^{0.2}}\right)$$
$$= -1.007 \times 10^4 \text{ kJ}$$

The change in internal energy is

**Closed Thermodynamic System**

$$Q - W = \Delta U + \Delta KE + \Delta PE$$
$$\Delta U = Q - W$$
$$-5 \text{ kJ} - (-1.007 \times 10^4 \text{ kJ})$$
$$= 10\,065 \text{ kJ} \quad (10\,000 \text{ kJ})$$

**The answer is (D).**

**65.** Treat the base as a circular sector. The length of the arc is

**Circular Sector**

$$\phi = s/r$$
$$s = \phi r$$
$$= \left(\frac{10°}{360°}\right)(10 \text{ cm})\left(2\pi \frac{\text{rad}}{\text{deg}}\right)$$
$$= 1.745 \text{ cm}$$

The area of the base is

**Circular Sector**

$$A = sr/2$$
$$= \frac{(1.745 \text{ cm})(10 \text{ cm})}{2}$$
$$= 8.725 \text{ cm}^2$$

The volume of the solid is

$$V = Ah$$
$$= (8.725 \text{ cm}^2)(10 \text{ cm})$$
$$= 87.25 \text{ cm}^3 \quad (87 \text{ cm}^3)$$

**The answer is (C).**

**66.** The equation simplifies to

$$\frac{d}{dy}\left(\sqrt{(y^5)} + 2\sqrt{y}\right) = \frac{d}{dy}(y^{2.5} + 2y^{1/2})$$
$$= \frac{d}{dy}(y^{2.5}) + \frac{d}{dy}(2y^{1/2})$$
$$= 2.5y^{1.5} + y^{-1/2}$$
$$= \frac{1}{\sqrt{y}}(2.5y^2 + 1)$$

**The answer is (D).**

**67.** Let $v = x^2 + 1$ and $c$ stand for the constant of integration.

This means $dv = 2xdx$, and

$$\int \frac{2xdx}{x^2+1} = \int \frac{dv}{v} = \ln v + c$$

Inserting the definite integral limits,

$$\int \frac{2xdx}{x^2+1} = (\ln v + c)\Big|_{x=4}^{x=8} = (\ln(x^2+1) + c)\Big|_{x=4}^{x=8}$$
$$= \left(\ln((8)^2 + 1) + c\right) - \left(\ln((4)^2 + 1) + c\right)$$
$$= \ln\left(\frac{8^2+1}{4^2+1}\right)$$
$$= 1.3$$

**The answer is (A).**

**68.** For this equation, the value of $dF/dx$ is

$$\frac{dF}{dx} = 32x^3 - 192x^2 + 352x - 192$$
$$= (8)(4x^3 - 24x^2 + 44x - 24)$$

At a maxima or minima, $dF/dx = 0$.

Thus, at the maxima or minima,

$$\frac{dF}{dx} = (8)\big((x-1)(x-2)(x-3)\big) = 0$$

A maxima or minima occurs at $x = 1$ or $x = 2$ or $x = 3$.

At the maxima, $d^2F/dx^2 < 0$, so calculate the value of $d^2F/dx^2$ at $x = 1, 2,$ and $3$.

$$\frac{d^2F}{dx^2} = \frac{d^2F}{d(1)^2} = 8$$

$$\frac{d^2F}{dx^2} = \frac{d^2F}{d(2)^2} = -4$$

$$\frac{d^2F}{dx^2} = \frac{d^2F}{d(3)^2} = 8$$

The maxima occurs when $x = 2$.

**The answer is (C).**

**69.** This is a mass balance problem. The variable $m$ is mass of nickel over the course of the year.

inputs = outputs + change in storage
$m_{\text{purchase}} = m_{\text{product}} + m_{\text{scrap}} + m_{\text{landfill}} + (m_{\text{inv,final}} - m_{\text{inv,initial}})$

Solve for the mass of nickel landfilled.

$m_{\text{landfill}} = m_{\text{purchase}} - m_{\text{product}} - m_{\text{scrap}} - (m_{\text{inv,final}} - m_{\text{inv,initial}})$
$= 4 \text{ metric tons} - (30 \text{ metric tons})(8\%)$
$\quad -(3 \text{ metric tons})(8\%)$
$\quad -(1 \text{ metric ton} - 0.5 \text{ metric ton})$
$= 0.86 \text{ metric ton}$

**The answer is (A).**

**70.** Applying Newton's Raphson method,

$$f(x) = x^3 - 2.5x - 5 = 0$$

$$\frac{df}{dx} = 3x^2 - 2.5$$

$$X_{n+1} = X_n - \frac{f(X_n)}{f'(X_n)} = X_n - \left(\frac{(X_n)^3 - 2.5X_n - 5}{3X_n^2 - 2.5}\right)$$

If the value of $x$ is 2, the initial equation comes out to a value of

$$x^3 - 2.5x - 5 = (2)^3 - (2.5)(2) - 5$$
$$= -2$$

If the value of $x$ is 3, the initial equation comes out to a value of

$$x^3 - 2.5x - 5 = (3)^3 - (2.5)(3) - 5$$
$$= 14$$

Because there is a change from a negative value to a positive value between $x = 2$ and $x = 3$, it can be inferred that the solution lies between those values. Choose an initial value of $X_n = 3$ and calculate; then calculate the answer using the found value, then calculate the answer using the value found from that calculation, and so on, until the difference between successive iterations is less than 5%. Tabulate the results.

| $n$ | $X_n$ | $X_{n+1}$ |
|---|---|---|
| 0 | 3 | 2.396 |
| 1 | 2.396 | 2.285 |
| 2 | 2.285 | 2.28 (2.3) |

#### The answer is (B).

**71.** The experimental values, in numerical order, are

60%, 61%, 68%, 68%, 70%, 70%, 70%, 70%, 71%, 72%

The middle of that range is 70%. Since there are 10 numbers, the median lies in the middle between the fifth and sixth values.

#### The answer is (C).

**72.** The mean value is the average of all values, including all repeated values.

$$\frac{\begin{matrix}64\% + 68\% + 70\% \\ + 70\% + 72\% + 82\% \\ + 70\% + 70\% + 68\%\end{matrix}}{9} = 70.44\% \quad (70\%)$$

#### The answer is (C).

**73.** The mode is the most frequently occurring number in a data set, which in this case is 4%.

#### The answer is (C).

**74.** For the minimum energy to be drawn, the cycle must be a Carnot cycle. The equation for the ratio of work done by an engine to heat drawn from a heat reservoir in a Carnot cycle is

$$\frac{\text{work done by engine}}{\text{heat drawn from reservoir}} = \frac{T_H - T_c}{T_{|c|}}$$

The absolute temperature in the heat reservoir is

$$T_H = 120°C + 273° = 393K$$

The absolute temperature of the sink is

$$T_c = 25°C + 273° = 298K$$

For the engine to do 1000 kJ of work, the minimum heat required is

$$\text{heat drawn from reservoir} = \frac{\text{work done}}{T_H - T_c} T_{|c|}$$
$$= \left(\frac{1000 \text{ kJ}}{393 \text{ K} - 298 \text{ K}}\right)(393 \text{K})$$
$$= 4136.8 \text{ kJ} \quad (4100 \text{ kJ})$$

#### The answer is (D).

**75.** Consult a hazardous waste compatibility chart. [**Hazardous Waste Compatibility Chart**]

Sodium diethyldithiocarbamate is a dithiocarbamate, and ethanol is an alcohol. As can be seen in a compatibility chart, there are no consequences of combining alcohols and dithiocarbamates. A combination of aliphatic amines and dithiocarbamates has unknown consequences. Mixing aldehydes with dithiocarbamates may generate flammable and toxic gases. Mixing organic acids with dithiocarbamates may generate heat, as well as flammable and toxic gases.

#### The answer is (B).

**76.** The equation for the ionic equilibrium constant of barium sulfate is

$$K = \frac{\begin{matrix}(\text{concentration of barium ions})\\ \times (\text{concentration of sulfate ions})\end{matrix}}{i}$$

The concentration of barium ions is equal to the concentration of sulfate ions, so the ionic equilibrium constant of barium sulfate is

$$K = \frac{(1.012 \times 10^{-5})^2}{1}$$
$$= 1.024 \times 10^{-10} \text{ g·mol/L} \quad (1.02 \times 10^{-10} \text{ g·mol/L})$$

#### The answer is (B).

**77.** Find the cost of the initial investment after 10 years. The equation for the cost of the initial investment is

$$F_{\text{inv}} = P_{\text{inv}}\left(\frac{F}{P}\right)\Big|_{t=10\text{yr}, i=6\%}$$

From interest tables, the factor $F/P$ for $t = 10$ yr and $i = 6\%$ is 1.7908. [**Interest Rate Tables**]

The value of the initial investment after 10 years is

$$F_{inv} = P_{inv}\left(\frac{F}{P}\right)\bigg|_{t=10\text{yr}, i=6\%}$$
$$= (\$50,000,000)(1.7908)$$
$$= \$89,540,000$$

For the next 10 years, the net profit each year is

annual profit = value of sales − annual expenses
$$= \$15,000,000 - \$5,000,000$$
$$= \$10,000,000$$

The equation for final value of the net annual profit is

$$(\text{annual profit})\left(\frac{F}{A}\right)\bigg|_{10 \text{ years } 6\%}$$

From interest tables, the factor $F/A$ for $t = 10$ yr and $i = 6\%$ is 13.1808. [Interest Rate Tables]

The final value of the net annual profit is

$$(\text{annual profit})\left(\frac{F}{A}\right)\bigg|_{t=10\text{yr}, i=6\%} = (\$10,000,000)(13.1808)$$
$$= \$131,808,000$$

The net future value after 10 years is

$\$131,808,000 - \$89,540,000 = \$42,268,000 \quad (\$42,000,000)$

**The answer is (C).**

**78.** The equation for the velocity at the point is

$$V = \sqrt{\frac{2g(P_o - P_s)}{\gamma}}$$

$V$ is the velocity in units of m/s, $P_o$ is the stagnation pressure in units of N/m², $P_s$ is the static pressure in units of N/m², and $\gamma$ is the specific weight of the fluid, which is equal to the density of the fluid, $\rho$, times the gravitational acceleration, $g$. [**Density, Specific Volume, Specific Weight, and Specific Gravity**]

Find the difference between of the static pressure and the stagnation pressure. The stagnation head is known, and the position of the pitot tube can be used as the value of the static head. The difference between the two is 1.5 m − 0.8 m = 0.7 m. Convert to units of N/m³.

$$\left(\frac{0.7 \text{ m}}{\left(33.9 \frac{\text{ft}}{\text{atm}}\right)\left(0.3 \frac{\text{m}}{\text{ft}}\right)}\right)(1 \text{ atm})(1.01)(10)^5 = 6930 \text{ N/m}^3$$

Specific weight can be expressed as $\rho g$, which means the equation for velocity can be simplified to remove the gravitational acceleration. The density of water is $1 \times 10^3$ kg/m³. The velocity is

$$V = \sqrt{\frac{2g(P_o - P_s)}{\gamma}}$$
$$= \sqrt{\frac{2(P_o - P_s)}{\rho}}$$
$$= \sqrt{\frac{(2)\left(6930 \frac{\text{N}}{\text{m}^3}\right)}{1 \times 10^3 \frac{\text{kg}}{\text{m}^3}}}$$
$$= 3.72 \text{ m/s} \quad (3.7 \text{ m/s})$$

**The answer is (D).**

**79.** From Bernoulli's equation for two points on the surface, just before and after the change in the cross-sectional flow, if the pressure at the top of the liquid at both points is the same, it is the pressure of the prevailing atmosphere. Also, the velocity before the change is expected to be more due to the smaller cross section, since volumetric flow is constant at steady flow. Since there is no pump work and the friction loss can be considered negligible, the height of the liquid will increase to have a constant total head.

**The answer is (A).**

**80.** Two environmental experts can be selected out of six in $_6C_2$ ways.

$$\frac{6!}{2!(6-2)!} = \frac{6!}{2!\,4!}$$
$$= \frac{(6)(5)(4)(3)(2)(1)}{(2)(1)(4)(3)(2)(1)}$$
$$= 15$$

Three lay persons can be selected out of six in $_6C_3$ ways.

$$\frac{(6)!}{3!(6-3)!} = \frac{(6)(5)(4)(3)(2)(1)}{(3)(2)(1)(3)(2)(1)}$$
$$= 20$$

The total number of ways the committee can be formed is

$$(15)(20) = 300$$

**The answer is (C).**

**81.** This is a mass balance problem. The molecular weight of tetrahydrothiophene (THT, $C_4H_8S$) is

$$MW_{THT} = (4)\left(12\ \frac{g\ C}{mol}\right) + (8)\left(1\ \frac{g\ H}{mol}\right) + (1)\left(32\ \frac{g\ S}{mol}\right)$$

$$= 88\ g/mol$$

The mass fraction of THT that is sulfur is

$$\frac{\left(32\ \frac{g\ S}{mol}\right)}{\left(88\ \frac{g\ THT}{mol}\right)} = 32\ g\ S/88\ g\ THT$$

Every sulfur atom in the tetrahydrothiophene becomes a molecule of sulfur dioxide ($SO_2$) during combustion. The molecular weight of sulfur dioxide is

$$MW_{SO_2} = (2)\left(16\ \frac{g\ O}{mol}\right) + (1)\left(32\ \frac{g\ S}{mol}\right)$$

$$= 64\ g/mol$$

64 g/mol/32 g/mol = 2 g $SO_2$/g S, meaning that for every gram of sulfur entering combustion, 2 g of sulfur dioxide are created. The mass balance is

sulfur entering combustion = sulfur exiting combustion

$$= \frac{SO_2\ exiting\ combustion}{2\ \frac{g\ SO_2}{g\ S}}$$

Solve for the $SO_2$ in the post-combustion stream.

$$\frac{sulfur\ entering}{combustion} = \left(2\ \frac{g\ SO_2}{g\ S}\right)(sulfur\ entering\ combustion)$$

$$= \frac{\left(2\ \frac{g\ SO_2}{g\ S}\right)\left(\frac{32\ g\ S}{88\ g\ THT}\right)}{10^6\ \frac{mg}{kg}}$$

$$\times \left(40\ \frac{mg\ THT}{m^3\ natural\ gas}\right)$$

$$\times (120 \times 10^6\ m^3\ natural\ gas)$$

$$= 3500\ kg\ SO_2$$

**The answer is (B).**

**82.** The equation for the benefit-cost ratio is

$$B/C$$

$B$ is the value of the benefits and $C$ is the cost.

The benefit-cost ratio is

$$B/C = \frac{\$8,125,000}{\$5,000,000} = 1.625 \quad (1.6)$$

**The answer is (C).**

**83.** Find the present worth of the benefits. From interest rate tables, the factor $P/A$ for $t = 10$ yr and $i = 6\%$ is 7.3601. [**Interest Rate Tables**]

$$(\$3,000,000) \qquad = \$3,000,000$$
$$+ \left((\$3,000,000)\left(\frac{P}{A}\right)\bigg|_{t=10\ yr, i=6\%}\right) \quad +(\$3,000,000)(7.3601)$$
$$= \$25,080,000$$

The benefit-cost ratio is

$$B/C = \frac{\$25,080,000}{\$20,000,000} = 1.254 \quad (1.25)$$

**The answer is (C).**

**84.** Find the ratio of the future worth to the present cost.

$$\frac{F}{P} = \frac{\$200,000,000}{\$100,000,000} = 2$$

From the equation for future worth as a function of present cost, the rate of return is

**Engineering Economics: Factor Table**
$$F = (1+i)^n P$$
$$\frac{F}{P} = (1+i)^n$$
$$2 = (1+i)^{10}$$
$$i = 6.9\% \quad (7\%)$$

**The answer is (C).**

**85.** The ratio of the future worth to the annual investment is

$$\frac{F}{A} = \frac{\$60,000,000}{\$10,000,000} = 6$$

Search the interest rate tables for $\frac{F}{A} = 6$ at $n = 5$.

Interpolating between the 8% and 10% interest rate tables, for $n = 5$ yr and $F/A = 6$, $i \approx 9\%$. [**Interest Rate Tables**]

**The answer is (C).**

**86.** From a table of typical material properties, the density of steel is 7.8 $Mg/m^3$. [**Material Properties**]

The CPR is

$$\text{CPR} = \frac{\Delta W}{\rho A t} = \frac{(33 \text{ g})\left(\dfrac{1 \text{ Mg}}{10^6 \text{ g}}\right)}{\left(7.8 \dfrac{\text{Mg}}{\text{m}^3}\right)(2.65 \text{ in}^2)\left(0.0254 \dfrac{\text{m}}{\text{in}}\right)^2} \times \left(\dfrac{1 \text{ m}}{1000 \text{ mm}}\right)(1 \text{ yr})$$

$$= 2.5 \text{ mm/yr}$$

**The answer is (D).**

**87.** From a list of expected $LD_{50}$ values, the $LD_{50}$ of acetaminophen is 142 mg/kg of body weight, the dose given in the problem statement. [**Dose-Response Curves**]

$LD_{50}$ is defined as the median lethal single dose, based on laboratory tests, expected to kill 50% of a group of test animals, usually by oral or skin exposure. The number of expected deaths is

$$\frac{\text{number of}}{\text{expected deaths}} = \frac{\text{number of individuals receiving the } LD_{50} \text{ dose}}{2}$$

$$= \frac{58}{2}$$

$$= 29$$

**The answer is (C).**

**88.** Using the molecular weight of benzene and the molar volume of an ideal gas, convert the 4.6 ppb benzene to units of $\mu g/m^3$. The molecular weight of benzene is

$$\text{MW}_\text{benzene} = (6)\left(12 \dfrac{\text{g C}}{\text{mol}}\right) + (6)\left(1 \dfrac{\text{g H}}{\text{mol}}\right) = 78 \text{ g/mol}$$

$$4.6 \text{ ppb} = \frac{4.6 \text{ parts benzene}}{10^9 \text{ parts air}} = \frac{4.6 \text{ mol}}{10^9 \text{ mol}}$$

$$= \frac{\left(78 \dfrac{\text{g}}{\text{mol}}\right)\left(10^6 \dfrac{\mu g}{g}\right)(4.6 \text{ mol})}{(10^9 \text{ mol})\left(22.4 \dfrac{\text{L}}{\text{mol}}\right)\left(\dfrac{1 \text{ m}^3}{1000 \text{ L}}\right)}$$

$$= 16 \ \mu g/m^3$$

The excess cancer risk (i.e., the fraction of the population who will get cancer) due to inhaling this concentration of benzene is

inhalation risk = (concentration in air) × (inhalation slope factor)

$$= \left(16 \dfrac{\mu g}{m^3}\right)\left(\dfrac{5 \times 10^{-6}}{1 \dfrac{\mu g}{m^3}}\right)$$

$$= 8.0 \times 10^{-5}$$

The number of additional cases of cancer due to benzene in the population of 2 million is

additional cases of cancer = (population)(risk)
$$= (2 \times 10^6)(8.0 \times 10^{-5})$$
$$= 160$$

**The answer is (C).**

**89.** Use the EPA recommended values for estimating intake. [**Intake Rates—Variable Values**]

The standard soil ingestion rate for a seven-year-old child is 50 mg/d. Exposure duration appears in both the numerator and denominator, so it cancels out; for ease of calculation, an exposure duration of one year is chosen. The equation for the averaging time is

**Intake Rates—Variable Values**

$$AT = (ED)(365 \text{ days/year})$$

Convert the concentration to mg/kg.

$$CS = 5 \text{ ppm}$$
$$= \left(\dfrac{5 \text{ g glyphosate}}{10^6 \text{ g soil}}\right)\left(1000 \dfrac{\text{mg}}{\text{g}}\right)\left(1000 \dfrac{\text{g}}{\text{kg}}\right)$$
$$= 5 \text{ mg/kg}$$

The formula for calculating chronic daily intake for ingestion of a chemical in soil is

Exposure

$$\text{CDI} = \frac{(\text{CS})(\text{IR})(\text{CF})(\text{FI})(\text{EF})(\text{ED})}{(\text{BW})(\text{AT})}$$

$$= \frac{\left(5\ \frac{\text{mg}}{\text{kg}}\right)\left(50\ \frac{\text{mg}}{\text{d}}\right)\left(10^{-6}\ \frac{\text{kg}}{\text{mg}}\right)}{(33\ \text{kg})\left(\left(365\ \frac{\text{d}}{\text{y}}\right)(1\ \text{y})\right)}\left(10^6\ \frac{\text{ng}}{\text{mg}}\right)$$

$$= 7.6\ \text{ng/kg·d}$$

### The answer is (D).

**90.** Option C is an example of an additive effect, option B is an example of an antagonistic effect, and option D is an example of a synergistic effect. [**Dose-Response Curves**]

### The answer is (D).

**91.** Use the EPA-recommended values for estimating intake rates. [**Intake Rates—Variable Values**]

The standard amount of air breathed by a three-year-old child is 8.3 m$^3$/d, and the standard average body weight of a three-year-old child is 16 kg. The equation for the averaging time is

**Intake Rates—Variable Values**
$$\text{AT} = (\text{ED})(365\ \text{days/year})$$

The exposure time is 24 h/d. Exposure duration appears in both the numerator and denominator, so it cancels out; for ease of calculation, an exposure duration of one year is chosen. Convert units of concentration to mg/m$^3$.

$$\text{CW} = \left(0.02\ \frac{\mu\text{g lead}}{\text{m}^3}\right)\left(\frac{1\ \text{mg}}{1000\ \mu\text{g}}\right) = 2.0 \times 10^{-5}\ \text{mg/m}^3$$

Convert amount of air breathed to m$^3$/h.

$$\text{IR} = \left(\frac{8.3\ \text{m}^3}{\text{d}}\right)\left(\frac{\text{d}}{24\ \text{h}}\right) = 0.35\ \text{m}^3/\text{h}$$

The formula for calculating chronic daily intake for inhaled chemicals is

Exposure

$$\text{CDI} = \frac{(\text{CA})(\text{IR})(\text{ET})(\text{EF})(\text{ED})}{(\text{BW})(\text{AT})}$$

$$= \frac{\left(2.0 \times 10^{-5}\ \frac{\text{mg}}{\text{m}^3}\right)\left(0.35\ \frac{\text{m}^3}{\text{h}}\right)}{(16\ \text{kg})\left(\left(365\ \frac{\text{d}}{\text{y}}\right)(1\ \text{y})\right)}\left(10^6\ \frac{\text{ng}}{\text{mg}}\right)$$

$$\times \left(24\ \frac{\text{h}}{\text{d}}\right)\left(365\ \frac{\text{d}}{\text{y}}\right)(1\ \text{y})$$

$$= 11\ \text{ng/kg·d}$$

### The answer is (B).

**92.** When using the rational formula to calculate peak runoff, the time of concentration is defined as the time required for water to travel from the most hydraulically remote point in a watershed to the point of interest. The rainfall for the time of concentration for a given storm duration is used as the rainfall intensity in the rational formula. The intensity of rainfall for the time of concentration can be linearly interpolated from the intensity of rainfall at 15 minutes and the intensity of rainfall at 30 minutes.

$$I_{21} = I_{15} + \frac{d_{21} - d_{15}}{d_{30} - d_{15}}(I_{30} - I_{15})$$

$$= 8.84\ \frac{\text{in}}{\text{hr}} + \left(\frac{21\ \text{min} - 15\ \text{min}}{30\ \text{min} - 15\ \text{min}}\right)\left(\frac{6.06\ \text{in}}{\text{hr}} - \frac{8.84\ \text{in}}{\text{hr}}\right)$$

$$= 7.7\ \text{in/hr}$$

### The answer is (C).

**93.** Substitute the equation for area into the equation for net loss and solve to find the net loss per day during the winter.

$$L = \left(4.3 \times 10^{-6}\ \frac{\text{km}}{\text{d}}\right)A$$

$$= \left(4.3 \times 10^{-6}\ \frac{\text{km}}{\text{d}}\right)\left(\left(\frac{2.3}{\text{km}^4}\right)V^2 + 3.1\ \text{km}^2\right)$$

$$= \left(4.3 \times 10^{-6}\ \frac{\text{km}}{\text{d}}\right)\left(\left(\frac{2.3}{\text{km}^4}\right)(9\ \text{km}^3)^2 + 3.1\ \text{km}^2\right)$$

$$= 8.1 \times 10^{-4}\ \text{km}^3/\text{d}$$

The net loss over the course of the winter is

$$L_{winter} = \left(\frac{365 \text{ d}}{4}\right) L$$

$$= \left(\frac{365 \text{ d}}{4}\right)\left(8.1 \times 10^{-4} \frac{\text{km}^3}{\text{d}}\right)$$

$$= 0.074 \text{ km}^3$$

**The answer is (C).**

**94.** This is a mass (volume) balance problem. The equation for the water held in the reservoir at the end of each hour is

water held in reservoir
at end of this hour = water held in reservoir
at end of previous hour
+inflow during this hour
−outflow during this hour

Calculate the water held for each hour.

| time interval (h) | inflow (m³/h) | outflow (m³/h) | water held at end of time interval (m³) |
|---|---|---|---|
| 1 | 500 | 0 | 500 |
| 2 | 4500 | 1000 | 4000 |
| 3 | 5000 | 3500 | 5500 |
| 4 | 4500 | 4000 | 6000 |
| 5 | 2500 | 3500 | 5000 |
| 6 | 1500 | 3000 | 3500 |
| 7 | 1000 | 2500 | 2000 |
| 8 | 500 | 1500 | 1000 |
| 9 | 500 | 1000 | 500 |
| 10 | 500 | 1000 | 0 |
| 11 | 500 | 500 | 0 |
| 12 | 500 | 500 | 0 |

The peak volume is held in the reservoir during the 4 h time interval. This can be deduced from looking at the chart in the problem statement without calculating the water held, because after 4 h the outflow exceeds the inflow.

**The answer is (C).**

**95.** The slope is 5%, so $m = 0.5$. Slope is rise over run, so the angle of the slope can be determined from the percentage slope.

$$\theta = \tan^{-1}(5\%) = \tan^{-1}(0.05)$$

The topographic factor for this lot is

$$L_S = \left(\frac{\lambda}{72.6}\right)^m (65.41 \sin^2\theta + 4.56 \sin\theta + 0.065)$$

$$= \left(\frac{150 \text{ ft}}{72.6}\right)^{0.5} \binom{65.41 \sin^2(\tan^{-1}(0.05))}{+4.56 \sin(\tan^{-1}(0.05)) + 0.065}$$

$$= 0.66$$

**The answer is (A).**

**96.** The density of water is 997 kg/m³. [**Selected Liquids and Solids**]

A gradient of 5% converts to a gradient of 1 m/20 m (1 m/20 m = 5%). The shear stress at maximum depth in the channel is

$$\tau_d = \gamma dS$$

$$= \left(997 \frac{\text{kg}}{\text{m}^3}\right)(1 \text{ m})\left(\frac{1 \text{ m}}{20 \text{ m}}\right)$$

$$= 50 \text{ kg/m}^2$$

**The answer is (D).**

**97.** The formula for the volumetric loading rate is

**Activated Sludge**

Organic loading rate (volumetric) = $Q_0 S_0$

The factor for converting mg/L to lbf/MG is 8.34 (lbf/MG)/(mg/L). [**Mass Calculations**]

$$\begin{aligned}\text{Organic loading rate} \\ \text{(volumetric)}\end{aligned} = Q_0 S_0$$

$$= (1.5 \text{ MGD})\left(130 \frac{\text{mg}}{\text{L}}\right)\left(8.34 \frac{\frac{\text{lbf}}{\text{MG}}}{\frac{\text{mg}}{\text{L}}}\right)$$

$$= 1600 \text{ lbf BOD/d}$$

**The answer is (C).**

**98.** The removal requirements for *Giardia* and viruses are 3-log (99.9%) and 4-log (99.99%), respectively. [**Disinfection**]

Conventional treatment typically removes 2.5-log (99.7%) of *Giardia* and 2-log (99%) of viruses. For both *Giardia* and viruses, conventional treatment removes less *Giardia* and viruses than required.

**The answer is (D).**

**99.** Kinematic viscosity is a function of absolute viscosity and density.

**Fluid Mechanics: Definitions**

$$\nu = \mu/\rho$$

From a table of properties of water, the kinematic viscosity of water at 20°C is 0.000001003 m²/s. [**Properties of Water (SI Metric Units)**]

The acceleration of gravity is 9.807 m/s². [**Fundamental Constants**]

The Carmen-Kozeny equation for head loss in mono-sized media is

**Filtration Equations: Carmen-Kozeny Equation**

$$h_f = \frac{f'L(1-\eta)v_s^2}{\eta^3 gd}$$

The equation for the friction factor is

**Filtration Equations**

$$f' = 150\left(\frac{1-\eta}{\text{Re}}\right) + 1.75$$

The equation for the Reynolds number is

**Filtration Equations**

$$\text{Re} = \frac{v_s \rho d}{\mu}$$

Combine the equation for kinematic viscosity with the equation for the Reynolds number to calculate the Reynolds number in terms of kinematic viscosity.

$$\text{Re} = \frac{v_s \rho d}{\mu} = \frac{v_s d}{\nu}$$

$$f' = 150\left(\frac{1-\eta}{\text{Re}}\right) + 1.75 = \frac{150\nu}{v_s d}(1-\eta) + 1.75$$

$$h_f = \frac{f'L(1-\eta)v_s^2}{\eta^3 gd} = \frac{\left(\frac{150\nu}{v_s d}(1-\eta) + 1.75\right)L(1-\eta)v_s^2}{\eta^3 gd}$$

$$= \frac{\left[\left(\frac{(150)\left(0.000001003 \frac{\text{m}^2}{\text{s}}\right)}{\left(0.003 \frac{\text{m}}{\text{s}}\right)(1 \text{ mm})\left(\frac{1 \text{ m}}{1000 \text{ mm}}\right)}\right)(1-0.4) + 1.75\right]}{(0.4)^3\left(9.807 \frac{\text{m}}{\text{s}^2}\right)(1 \text{ mm})\left(\frac{1 \text{ m}}{1000 \text{ mm}}\right)}$$

$$\times (0.5 \text{ m})(1-0.4)\left(0.003 \frac{\text{m}}{\text{s}}\right)^2$$

$$= 0.14 \text{ m}$$

Note that the italicized lowercase letter "v" (for velocity) and the italicized Greek lowercase letter nu ($\nu$, for kinematic viscosity) bear an unfortunate degree of resemblance to each other and should not be confused.

*The answer is (B).*

**100.** Use Theim's equation for drawdown of a confined aquifer.

**Thiem Equation**

$$Q = \frac{2\pi T(h_2 - h_1)}{\ln\left(\frac{r_2}{r_1}\right)}$$

This equation shows that if the ratio of $r_1$ to $r_2$ and $T$ are the same, the difference in drawdown at $r_1$ and $r_2$ is directly proportional to the pumping rate. In other words, if the aquifers and well construction are the same,

$$\frac{Q_{\text{well 2}}}{Q_{\text{well 1}}} = \frac{\dfrac{2\pi T_{\text{well 2}}(h_2 - h_1)_{\text{well 2}}}{\ln\left(\dfrac{r_2}{r_1}\right)_{\text{well 2}}}}{\dfrac{2\pi T_{\text{well 1}}(h_2 - h_1)_{\text{well 1}}}{\ln\left(\dfrac{r_2}{r_1}\right)_{\text{well 1}}}} = \frac{(h_2 - h_1)_{\text{well \#2}}}{(h_2 - h_1)_{\text{well \#1}}}$$

Solving for $(h_2 - h_1)_{\text{well 2}}$ yields

$$(h_2 - h_1)_{\text{well 2}} = \frac{Q_{\text{well 2}}(h_2 - h_1)_{\text{well 1}}}{Q_{\text{well 1}}}$$

$$= \frac{\left(46 \frac{\text{L}}{\text{s}}\right)(0.14 \text{ m})}{\left(28 \frac{\text{L}}{\text{s}}\right)}$$

$$= 0.23 \text{ m}$$

*The answer is (C).*

**101.** The storativity is

$$S = S_s b = (12 \text{ m})\left(\frac{0.00032}{\text{m}}\right)$$

$$= 0.0038$$

*The answer is (B).*

**102.** The formula for the area of the sides of a cylinder of height $h$ and radius $r$ is

$$A = 2\pi rh$$

This is the area across which radial flow of water towards the well occurs in the aquifer. Darcy's law is

**Darcy's Law**
$$Q = -KA(dh/dx)$$

In the case of a well, radial flow is toward the well, so $x = r$ and $dh/dx$ is $-dh/dr$. Darcy's law becomes

$$Q = KA\frac{dh}{dr}$$

For an unconfined aquifer, substitute the formula for area of a cylinder of height $h$, separate the variables $h$ and $r$, and integrate across the interval given in the problem statement.

$$Q = K2\pi rh\frac{dh}{dr}$$

$$\int_{r_w}^{r}\frac{dr}{r} = \frac{2\pi K}{Q}\int_{h_w}^{h}hdh$$

$$\ln\frac{r}{r_w} = \frac{\pi K}{Q}(h^2 - h_w^2)$$

This is Dupuit's formula with $h_2 = h$ and $h_1 = h_w$. Solve for $h$. [**Dupuit's Formula**]

$$h = \sqrt{\frac{Q\ln\frac{r}{r_w}}{\pi K} + h_w^2}$$

$$= \sqrt{\frac{\left(32\,\frac{L}{s}\right)\left(\frac{m^3}{1000\,L}\right)\ln\left(\frac{15\,m}{0.15\,m}\right)}{\pi\left(0.028\,\frac{m}{s}\right)} + (7.5\,m)^2}$$

$$= 7.6\,m$$

**The answer is (B).**

**103.** A diagram of the system is shown.

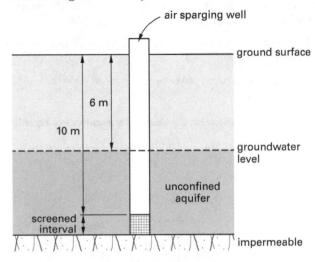

The formula for calculating density from specific gravity is

$$\rho = (SG)\rho_{water}$$

The pressure on a point at a vertical distance below the surface is

**Forces on Submerged Surfaces and the Center of Pressure**
$$P = P_{atm} + \rho gh$$

In this case, the sparging pressure, or gage pressure, is sought, so the equation simplifies to

$$P = \rho gh$$

The soil particles take up a fraction of $(1-n)$ of the soil matrix, where $n$ is the porosity of the soil. Therefore, the density of the overall soil including the pore space is

$$\rho_s = (1-n)\rho_{soil\ particles}$$

Similarly, the overall density of the water in the aquifer is

$$\rho_w = n\rho_{water}$$

The total pressure at the top of the screened interval is the pressure from the soil and the water combined. The density of water is 997 kg/m³. [**Selected Liquids and Solids**]

$$P_{tot} = P_s + P_w = \rho_s g h_s + \rho_w g h_w$$
$$= (1-n)\rho_{\text{soil particles}} g h_s + n\rho_{\text{water}} g h_w$$
$$= (1-n)(\text{SG}_{\text{soil particles}})\rho_{\text{water}} g h_s + n\rho_{\text{water}} g h_w$$
$$= \rho_{\text{water}} g\big((1-n)(\text{SG}_{\text{soil particles}})h_s + n h_w\big)$$
$$= \left(997\ \frac{\text{kg}}{\text{m}^3}\right)\left(9.8\ \frac{\text{m}}{\text{s}^2}\right)\binom{(1-0.4)(2.7)(10\ \text{m})}{+(0.4)(10\ \text{m}-6\ \text{m})}$$
$$\times\left(\frac{1000\ \text{kPa}}{1\ \text{Pa}}\right)$$
$$= 170\ \text{kPa}$$

The safety factor is 75%, so the design pressure is

$$(170\ \text{kPa})(75\%) = 130\ \text{kPa}$$

**The answer is (A).**

**104.** Storativity is the change in volume of water per unit area per unit change in head. [**Hydrology/Water Resources**]

The equation for storativity is

$$S = \frac{\Delta V}{A\Delta h}$$

Solve for $\Delta h$, the change in head.

$$\Delta h = \frac{\Delta V}{SA}$$
$$= \left(\frac{0.093\ \text{km}^3}{(0.14)(400\ \text{km}^2)}\right)\left(1000\ \frac{\text{m}}{\text{km}}\right)$$
$$= 1.7\ \text{m}$$

**The answer is (D).**

**105.** The equation for the Langmuir isotherm is

**Activated Carbon Adsorption: Langmuir Isotherm**

$$\frac{x}{m} = X = \frac{aKC_e}{1+KC_e}$$

The variables $x$ and $m$ can be flow rates. $C_e$ is the desired concentration of the contaminant after treatment with carbon adsorption. Solve for $x$.

$$x = \frac{maKC_e}{1+KC_e}$$

A mass balance for contaminant around the carbon adsorption unit yields

contaminant in = contaminant out + contaminant adsorbed

$$C_{in}Q_{air} = C_{out}Q_{air} + x$$

Substitute for $x$, solve for $m$, and substitute values.

$$C_{in}Q_{air} = C_{out}Q_{air} + \frac{maKC_e}{1+KC_e}$$

$$m = \frac{(C_{in}Q_{air} - C_{out}Q_{air})(1+KC_e)}{aKC_e}$$

$$= \frac{Q_{air}(C_{in}-C_{out})(1+KC_e)}{aKC_e}$$

$$= \frac{\left(0.5\ \frac{\text{m}^3}{\text{min}}\right)\left(60\ \frac{\text{min}}{\text{h}}\right)\left(24\ \frac{\text{h}}{\text{d}}\right)}{\left(0.33\ \frac{\text{g}}{\text{g}}\right)\left(0.2\ \frac{\text{m}^3}{\text{g}}\right)\left(2.4\ \frac{\text{g}}{\text{m}^3}\right)\left(1000\ \frac{\text{g}}{\text{kg}}\right)}$$
$$\times\left(\left(10\ \frac{\text{mg}}{\text{L}}\right)\left(\frac{1\ \text{g}}{1000\ \text{mg}}\right)\left(1000\ \frac{\text{L}}{\text{m}^3}\right)-2.4\ \frac{\text{g}}{\text{m}^3}\right)$$
$$\times\left(1+\left(0.2\ \frac{\text{m}^3}{\text{g}}\right)\left(2.4\ \frac{\text{g}}{\text{m}^3}\right)\right)$$

$$= 51\ \text{kg/d}$$

**The answer is (D).**

**106.** From a table of equivalent weights, the equivalent weight of $Ca^{2+}$ is 20.0. [**Lime-Soda Softening Equations**]

The equivalent concentration of $Ca^{2+}$ is

$$\frac{78\ \frac{\text{mg}}{\text{L}}}{20.0} = 3.9\ \text{meq/L}$$

From a table of equivalent weights, the equivalent weight of $Mg^{2+}$ is 12.2. [**Lime-Soda Softening Equations**]

The equivalent concentration of $Mg^{2+}$ is

$$\frac{95\ \frac{\text{mg}}{\text{L}}}{12.2} = 7.787\ \text{meq/L}$$

The $Na^+$ does not add to the total hardness. The total hardness of the water sample is

$$3.9\ \frac{\text{meq}}{\text{L}} + 7.787\ \frac{\text{meq}}{\text{L}} = 11.687\ \text{meq/L}\quad(12.0\ \text{meq/L})$$

**The answer is (C).**

**107.** Non-carbonate hardness is equal to the total hardness minus the carbonate hardness.

First, find the total hardness. From a table of equivalent weights, the equivalent weight of $Ca2+$ is 20.0. [**Lime-Soda Softening Equations**]

The equivalent concentration is

$$\frac{90 \frac{\text{mg}}{\text{L}}}{20.0} = 4.5 \text{ meq/L}$$

From a table of common radicals in water, the equivalent weight of $Mg^{2+}$ is 12.2. The equivalent concentration is

$$\frac{92 \frac{\text{mg}}{\text{L}}}{12.2} = 7.541 \text{ meq/L}$$

The $Na^+$ does not add to the total hardness. The total hardness of the water sample is

$$4.5 \frac{\text{meq}}{\text{L}} + 7.541 \frac{\text{meq}}{\text{L}} = 12.041 \text{ meq/L}$$

The carbonate hardness in terms of $CaCO_3$ is the same as the alkalinity as $CaCO_3$, 82 mg/L. From a table of common radicals in water, the equivalent weight of $CaCO_3$ is 50.0.

The equivalent concentration is

$$\frac{82 \frac{\text{mg}}{\text{L}}}{50.0} = 1.64 \text{ meq/L}$$

The non-carbonate hardness is

$$\begin{aligned}\text{non-carbonate hardness} &= \text{total hardness} - \text{carbonate hardness} \\ &= 12.041 \frac{\text{meq}}{\text{L}} - 1.64 \frac{\text{meq}}{\text{L}} \\ &= 10.401 \text{ meq/L} \quad (10.4 \text{ meq/L})\end{aligned}$$

**The answer is (C).**

**108.** The total suspended solids (TSS) is equal to the total solids (TS) minus the total dissolved solids (TDS).

First, find the TS.

The total mass of the 200 mL sample of wastewater, after being heated to 103°C without being filtered, is

$$47.760 \text{ mg} - 47.510 \text{ mg} = 0.25 \text{ mg}$$

1 L of waste water would then yield TS of

$$TS = \left(\frac{0.25 \text{ mg}}{200 \text{ mL}}\right)\left(1000 \frac{\text{mL}}{\text{L}}\right) = 1.25 \text{ mg/L}$$

The TDS is the residue left after the 400 mL sample is filtered, then heated to 600°C and allowed to cool.

$$47.382 \text{ mg} - 47.132 \text{ mg} = 0.25 \text{ mg}$$

The TDS is

$$TDS = \left(\frac{0.25 \text{ mg}}{400 \text{ mL}}\right)\left(1000 \frac{\text{mL}}{\text{L}}\right) = 0.625 \text{ mg/L}$$

The TSS is

$$\begin{aligned}TSS &= TS - TDS \\ &= 1.25 \frac{\text{mg}}{\text{L}} - 0.625 \frac{\text{mg}}{\text{L}} \\ &= 0.625 \text{ mg/L} \quad (0.63 \text{ mg/L})\end{aligned}$$

**The answer is (C).**

**109.** Use the variables for the population in the town and the metropolitan area shown in the figure. $P_0$ is the population of the town in the present, $P_1$ is the population of the town 10 years ago, $P_2$ is the population of the town 10 years in the future, $P_{R0}$ is the population of the metropolitan area in the present, $P_{1R}$ is the population of the metropolitan area 10 years ago, and $P_{2R}$ is the population of the metropolitan area 10 years in the future.

```
                    P₀
          P₁        P_R0              P₂
 P_1R ────┼──────────┼───────────────┼──── P_2R
        t = -10    t = 0          t = +10
```

The initial per capita water consumption is 169 gpcd. The per capita consumption after 10 years is

$$(169 \text{ gpcd})(1.01) = 170.69 \text{ gpcd}$$

Find the growth ratio constant.

For the past 10 years, $P_{R0}/P_{1R} = 1.20$. Assuming this holds constant, $P_{2R}/P_{R0} = 1.20$, and in turn $P_2/P_0 = 1.20$.

Find the total water consumption in the town after 10 years as a function of the population.

$$\begin{aligned}P_2(170.69 \text{ gpcd}) &= 1.20 P_0 (170.69 \text{ gpcd}) \\ &= P_0 (204.82 \text{ gpcd})\end{aligned}$$

The present water consumption in the town as a function of the population is $P_0(169 \text{ gcpd})$.

The increase in the total water consumption after 10 years is

$$\frac{P_0(204.82 \text{ gpcd}) - P_0(169 \text{ gpcd})}{P_0(169 \text{ gpcd})} \times 100\% = 21.2\% \quad (21\%)$$

**The answer is (B).**

**110.** An oxidizing environment is a necessary condition for the formation of iron oxide (rust) on iron.

**The answer is (C).**

# Practice Exam 2

## PROBLEMS

**1.** A waste sample is considered corrosive under the Resource Conservation and Recovery Act (RCRA) if it

- (A) corrodes a sample of iron at 25°C
- (B) is aqueous and has a pH of either less than 2 or greater than 12.5
- (C) contains ions of copper or nickel
- (D) contains hydroxyl ions

**2.** Which of the following is used to accelerate the sedimentation of sewage?

- (A) nitric acid
- (B) lime
- (C) potassium permanganate
- (D) chlorine

**3.** The average velocity of water at 25°C in the channel shown is 0.15 m/s.

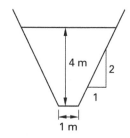

The mass flow rate is most nearly

- (A) 100 kg/s
- (B) 1800 kg/s
- (C) 2800 kg/s
- (D) 3800 kg/s

**4.** A closed tank contains water at 20°C. A manometer at the side of the tank contains fluid with a specific gravity of 5.6. If the barometric pressure is 750 mm Hg and the height of fluid in the manometer on the tank side is 10 cm lower than that on the air side, the pressure in the tank is most nearly

- (A) 4.5 kPa
- (B) 99 kPa
- (C) 100 kPa
- (D) 150 kPa

**5.** The rate constant for a coagulation reaction is known to be 50/d. The incoming concentration is 150 mg/L, and it is desired that the mixing of chemicals into the water for the coagulation produce a 90% reduction in concentration in this fully mixed reactor. The volume of the reactor is 5 L. Therefore, the flow rate that should be used is most nearly

- (A) 19 cm³/min
- (B) 50 cm³/min
- (C) 150 cm³/min
- (D) 500 cm³/min

**6.** A sharp-edged meter used to measure the flow of water at 25°C in a pipe of inner diameter 0.051 m registers a pressure difference of 10 Pa. The diameter of the orifice is 0.046 m. The flow rate in the pipe is most nearly

- (A) 0.21 m³/h
- (B) 0.32 m³/h
- (C) 0.43 m³/h
- (D) 0.51 m³/h

**7.** The data in the following table are for flow in the river shown at a given cross section.

| distance from the left bank, $d$ (m) | depth of river, $h$ (m) | velocity of flow, v (m/s) | |
|---|---|---|---|
| | | $0.3h$ | $0.7h$ |
| 1 | 1 | 0.5 | 0.5 |
| 2 | 3 | 0.6 | 0.3 |
| 3 | 4 | 0.8 | 0.6 |
| 4 | 5 | 0.9 | 0.6 |
| 5 | 5 | 0.9 | 0.6 |
| 6 | 5 | 0.9 | 0.6 |
| 7 | 5 | 1.0 | 0.8 |
| 8 | 2.5 | 0.6 | 0.4 |

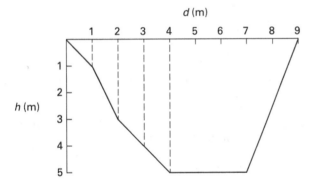

The average velocity in the river is most nearly

(A) 0.50 m/s

(B) 0.60 m/s

(C) 0.66 m/s

(D) 0.70 m/s

**8.** Toluene is released to a shallow pond. Its first-order removal rate constant is 0.067/h. The time it will take for the concentration of toluene in the pond to drop by one half is most nearly

(A) 1 h

(B) 10 h

(C) 20 h

(D) 30 h

**9.** A town has a population of 1.1 million in 1980, 1.2 million in 1990, and 1.3 million in 2000. The per-capita consumption of water was 45 L/person/d in 1980, 49 L/person/d in 1990, and 47 L/person/d in 2000. The pump for the drinking water delivery system is capable of delivering $76 \times 10^6$ L/d. The year in which there will be a need for additional capacity is most likely

(A) 2010

(B) 2020

(C) 2030

(D) 2040

**10.** The capacity of a town's pumping system is $110 \times 10^6$ L/d. The present population of the town is 1.5 million, and the average consumption of water is expected to remain constant at 36 L/person/d. The population a decade ago was 1.3 million, and a decade before that, the population was 1.2 million. The year in which the capacity will be fully utilized is most likely

(A) 20 yr from now

(B) 29 yr from now

(C) 73 yr from now

(D) 97 yr from now

**11.** The storage coefficient, $S$, of a confined aquifer can be given by

$$S = \frac{\eta \gamma_w b g}{E_w B}$$

In this equation, $\eta$ is porosity, $\gamma_w$ is the density of water, $b$ is the aquifer thickness, $E_w$ is the bulk modulus of compressibility of water ($2.07 \times 10^9$ N/m$^2$), $g$ is the gravitational constant, and $B$ is the barometric efficiency of the aquifer.

The storage coefficient of a confined aquifer can also be written as

$$S = \beta \gamma_w b g$$

In this equation, $\beta$ is the aquifer's compressibility, and $b$ and $\gamma_w$ are as before. The barometric efficiency of an aquifer whose compressibility is $1.9 \times 10^{-7}$ m$^2$/N and whose porosity is 0.2 is most nearly

(A) 0.000058

(B) 0.00051

(C) 0.51

(D) 0.13

**12.** A solution has a pH of 2. The mass in grams of solid sodium hydroxide (NaOH) needed to neutralize 1000 L of the solution is most nearly

(A) 0.4 g
(B) 10 g
(C) 400 g
(D) 2000 g

**13.** Water at 27°C flows through a 2 cm pipe at the rate of 1.0 cm³/s. The pressure drop per meter is most nearly

(A) 0.10 Pa/m
(B) 0.22 Pa/m
(C) 0.68 Pa/m
(D) 0.88 Pa/m

**14.** The retardation factor of a contaminant in an aquifer is 5. Therefore,

(A) four-fifths of the contaminant is absorbed onto the solids
(B) four-fifths of the contaminant is dissolved in the groundwater
(C) the groundwater travels at five times the speed at which the contaminant travels
(D) the contaminant is equally distributed between the solids and the groundwater

**15.** The initial volume of solid waste is 18.5 m³. The process of baling carried out before disposal reduces the volume by 85%. The compaction ratio is most nearly

(A) 5.7
(B) 6.7
(C) 8.5
(D) 19

**16.** Water at 20°C is pumped to the open top of an absorption system from an open tank. The height of water in the tank is 20 m. The height of the absorption system is 100 m. The total head loss, $h_f$, in the entire system is 1.0 m. The flow rate in the 4 cm pipe is 1 L/s. If the efficiency of the pump is 60%, the power of the pump is most nearly

(A) 0.90 kW
(B) 1.0 kW
(C) 1.3 kW
(D) 1.5 kW

**17.** A discharge wetland

(A) drains into a groundwater system
(B) receives water from a groundwater system
(C) does not exchange water with a groundwater system
(D) is also known as a spring

**18.** The pressure at the inlet of a centrifugal pump is 69 kPa. The pump generates 11 kW of power and is 48% efficient. The flow rate of water at 25°C is 0.1 L/s. The pressure at the outlet of the pump is most nearly

(A) $1.2 \times 10^5$ Pa
(B) $2.2 \times 10^5$ Pa
(C) $3.5 \times 10^5$ Pa
(D) $5.3 \times 10^7$ Pa

**19.** A 10 m long, 4 cm diameter cast iron pipe gradually contracts into a 10 m long, 2 cm cast iron pipe. The flow rate through the pipes is 10 L/min of water at 20°C. The equivalent length of a cast iron 4 cm pipe for the given flow configuration is most nearly

(A) 20 m
(B) 40 m
(C) 100 m
(D) 350 m

**20.** The residence time of a kiln must be doubled in order to meet new clean air regulations. This can be accomplished by

(A) doubling the rotational speed while leaving the other parameters constant
(B) halving the rotational speed while leaving the other parameters constant
(C) doubling the kiln rake while leaving the other parameters constant
(D) halving the kiln rake and doubling the rotational speed while leaving the other parameters constant

**21.** The average fugitive emission factor for valves in gas service at a chemical manufacturing plant is 0.0268 kg/h/valve. The equation for applying the emission factor is

$$E_{TOC} = F_A WF_{TOC} N$$

In this equation, $E_{TOC}$ is the emission rate, $F_A$ is the applicable average emission factor, $WF_{TOC}$ is the average weight fraction of total organic compounds in the stream, and $N$ is the number of pieces of equipment. A gas stream at a chemical manufacturing plant has 110 valves. The gas stream contains an average of 10% water vapor, with the remainder being organic compounds. Assume the facility operates 24 h/d, 350 d/yr. Using the emission factor approach, the estimated fugitive emissions for this gas stream are most nearly

(A) 2.95 kg/yr
(B) 2480 kg/yr
(C) 22 300 kg/yr
(D) 23 200 kg/yr

**22.** In the air over a rural area, sulfur dioxide is present at a partial pressure of $5 \times 10^{-9}$ atm. The temperature, $T$, is 20°C, and the pressure, $P$, is 1 atm. The number of moles of sulfur dioxide in 1 m³ of this air is most nearly

(A) $2 \times 10^{-10}$ mol
(B) $2 \times 10^{-7}$ mol
(C) $3 \times 10^{-6}$ mol
(D) $3 \times 10^{-5}$ mol

**23.** The concentration of carbon dioxide in the troposphere is 350 ppm (volume). The molar concentration of carbon dioxide in air at standard temperature and pressure is most nearly

(A) $2.4 \times 10^{-5}$ mol/L
(B) $1.6 \times 10^{-5}$ mol/L
(C) $3.5 \times 10^{-4}$ mol/L
(D) $7.8 \times 10^{-3}$ mol/L

**24.** A pumping system contains a centrifugal pump with an impeller diameter of 1 ft. When operating at 1000 rpm, the discharge head is 100 ft and the discharge flow rate is 10 gal/min. The specific speed of the pump is most nearly

(A) 100
(B) 200
(C) 300
(D) 400

**25.** The total organic carbon (TOC) of a compound is the mass of carbon in the compound divided by the total mass of the compound. The molecular formula of acetone is $C_3H_6O$. The TOC of acetone is most nearly

(A) 0.38 g carbon/g acetone
(B) 0.45 g carbon/g acetone
(C) 0.53 g carbon/g acetone
(D) 0.62 g carbon/g acetone

**26.** A centrifugal pump with a 1 m impeller diameter generates an effective head of 100 m at 1000 rpm. For a similar pump generating an effective head of 2000 m at 1000 rpm, the required impeller diameter would most nearly be

(A) 4.5 m
(B) 5.1 m
(C) 6.4 m
(D) 9.6 m

**27.** A sample of diluted wastewater (diluted 10 times) has a 5-day biological oxygen demand (BOD) of 5 mg/L. If the rate constant is 0.1/d, assuming a first-order chemical reaction, the ultimate BOD of the original wastewater is most nearly

(A) 0.50 mg/L
(B) 5.0 mg/L
(C) 13 mg/L
(D) 130 mg/L

**28.** The microorganisms that are used in secondary treatment of wastewaters with heavy carbonaceous organic dissolved matter can be best described as

(A) carbonaceous
(B) anaerobic
(C) aerobic
(D) nitrogenous

**29.** A city with a population of 19,000 is planning to modify its old sewer system. The ratio of the peak flow rate to the average flow rate is most nearly

(A) 1.1
(B) 1.9
(C) 2.7
(D) 5.0

**30.** An aerobic digester has an inflow of 2.8 m³/d. The influent 5-day BOD is 200 mg/L, while the influent suspended solids are 5 mg/L. The digester suspended solids are 10 mg/L. The reaction-rate constant is 0.50/d. The volatile fraction is 0.1. The solids retention time is 2 d. The fraction of influent 5-day BOD consisting of raw sewage is 0.5. Under these conditions, the volume of the aerobic digester is most nearly

(A) 5.6 m³
(B) 28 m³
(C) 34 m³
(D) 53 m³

**31.** The electrodialysis of a 0.1 N NaCl solution is carried out in 100 cells. The flow rate of the solution is 1 L/s, the removal efficiency is 60%, and the electrical efficiency is 98%. The current required for operation is most nearly

(A) 60 A
(B) 200 A
(C) 500 A
(D) 1000 A

**32.** During the secondary treatment of wastewater, it is known that a biochemical reaction takes place. In the language of chemistry, the reaction that affects wastewater can be best described as

(A) acidification
(B) hydration by adding water
(C) carbonation by adding carbon dioxide
(D) oxidation

**33.** The filter cake obtained by filtering settled sludge in a steeling tank is fed to a dryer. The dried product from the dryer contains 25% water. The feed to the dryer contains 63% water. If 100 kg of water is removed from the dryer per hour, the amount of feed to the dryer is most nearly

(A) 50 kg
(B) 100 kg
(C) 200 kg
(D) 400 kg

**34.** The disinfection of water is carried out in a stirred tank. The treatment process has water as the input at a rate of 38 000 L/h. The tank carries 3800 L of water at all times. Sufficient chlorine is fed to the water. The rate of disinfection can be described as a first-order reaction with a rate constant of 0.055/min. The efficiency of the system can be expected to be

(A) 0.050
(B) 0.25
(C) 0.50
(D) 0.75

**35.** In a sewer pipe with a 20 cm diameter, sewage flows to a depth of 15 cm. The hydraulic radius, $R$, of the flow configuration is most nearly

(A) 6 cm
(B) 8 cm
(C) 10 cm
(D) 12 cm

**36.** The dimensionless incinerability index, $I$, of a material in a waste stream is given by

$$I = C + \frac{100 \frac{\text{kcal}}{\text{g}}}{H}$$

In a particular waste stream, the mass percent of chlordane, $C$, is 0.02, and the heating value, $H$, is 2.71 kcal/g. The uncertainty in $H$ is $\pm 0.06$ kcal/g, and the uncertainty in $C$ is $\pm 10\%$. The uncertainty in $I$ is most nearly

(A) 0.06
(B) 0.20
(C) 0.30
(D) 0.80

**37.** The half-life of a biologically degraded contaminant is 1 h. Therefore, the kinetic constant of the reaction is most nearly

(A) 17/d
(B) 24/d
(C) 73/d
(D) 96/d

**38.** The standard deviation of a sample of data points is 45 ppm, and the sample average is 300 ppm. If a confidence level of 90% is sought, the probability of making a

Type II error is 10%, and five samples are taken, the minimum detectable relative difference is most nearly

(A) 5%

(B) 10%

(C) 20%

(D) 30%

**39.** Which of the following describes the consequence(s) of mixing cyanides with organic acids?

(A) heat is generated

(B) toxic and flammable gases are generated

(C) toxic substances are solubilized

(D) there are no consequences

**40.** A clay liner for a waste pond is 0.7 m thick. The effective porosity is 0.2, and the coefficient of permeability is $3 \times 10^{-7}$ cm/s. The hydraulic head at the liner is 2 m. The break-through time for leachate to penetrate the liner is

(A) 1.4 d

(B) 14 d

(C) 140 d

(D) leachate will not penetrate

**41.** The bioconcentration factor for benzo(a)pyrene in daphnia is approximately 10 000. The concentration of benzo(a)pyrene in daphnia living in water that has a benzo(a)pyrene concentration of 2 ppb (mass) is most nearly

(A) 20 mg/L

(B) 200 mg/L

(C) 2000 mg/L

(D) 20 000 mg/L

**42.** The economic criterion for wastewater reuse decisions is the

(A) cost of treatment processes for reuse versus cost of treatment processes for general water supply

(B) cost of conveying treated wastewater versus cost of conveying general water supply

(C) cost of treating and conveying wastewater versus cost of treating and conveying water

(D) cost of treating wastewater for use in irrigation, which is small, so that one can assume that wastewater reuse is always the most economical decision

**43.** The collection efficiency of fly ash particles of diameter 0.3 μm to 0.7 μm in an existing 600 m² electrostatic precipitator is approximately 92%. Collection efficiency must be increased to 96%. To achieve the higher efficiency, the area of the electrostatic precipitator will be increased. The new area is most nearly

(A) 630 m²

(B) 770 m²

(C) 1200 m²

(D) 1800 m²

**44.** A pulse jet/felt baghouse is to be used to trap quartz dust. The volumetric flow rate of the air to be treated is 2000 m³/min. The area of felt needed is most nearly

(A) 740 m²

(B) 1800 m²

(C) 2200 m²

(D) 5400 m²

**45.** A stream has an ultimate BOD, $S_s$, of 3 mg/L. Its flow rate, $Q_s$, is 3 m³/s. It receives a wastewater stream whose flow rate, $Q_w$, is 1 m³/s. The wastewater stream has an ultimate BOD, $S_w$, of 30 mg/L. The BOD of the stream just below the discharge point, $S_0$, is most nearly

(A) 10 mg/L

(B) 20 mg/L

(C) 30 mg/L

(D) 40 mg/L

**46.** The efficiency of a cyclone increases as

(A) particle size decreases while everything else remains the same

(B) the velocity of gases through the inlet increases while everything else remains the same

(C) particle density decreases while everything else remains the same

(D) the length of the cone increases while everything else remains the same

**47.** In wastewater collection systems, manholes are used

(A) on wide streets where there is no change in pipe slope

(B) where there is a change of pipe slope

(C) where streets are narrow

(D) where stormwater comes into wastewater collection systems

**48.** A stack of height 50 m emits a plume whose rise is 10 m. On a cloudy night with the wind at 4 m/s, the maximum ground-level concentration downwind of the stack is most nearly what distance from the stack?

(A) 0.50 km

(B) 1.3 km

(C) 2.0 km

(D) 3.8 km

**49.** The gross precipitation on a watershed is 12 in, and the maximum basin retention is 1 in. The runoff is most nearly

(A) 0.85 in

(B) 0.92 in

(C) 1.3 in

(D) 11 in

**50.** On a clear night with a wind speed of 4 m/s, the concentration of a compound at ground level 200 m downwind of a stack is $1.5 \times 10^{-4}$ g/m³. The compound is emitted at a rate of 60 g/s from a stack whose effective stack height is 20 m. The wind speed 20 m above ground level is 7 m/s. Relative to this concentration of $1.5 \times 10^{-4}$ g/m³, the concentration at the same point on a clear summer day with the sun higher than 60° above the horizon and at the same wind speed is most nearly

(A) 13 times lower

(B) the same

(C) 17 times higher

(D) 29 times higher

**51.** An incinerator is used to destroy contaminated soil at a hazardous waste remediation site. In order to destroy 99.99% of the hazardous constituents in the soil, the temperature of the incinerator must be maintained at 1100°C. In order to maintain this temperature, a supplemental fuel for the incinerator is required. Excess air supplied is 30%, and no. 2 fuel oil is to be used. At 1100°C and 30% excess air, 15 200 kJ/L are available from no. 2 fuel oil. The heat deficiency of the stream to be incinerated is 4020 MJ/h.

The rate at which no. 2 fuel oil must be fed to the incinerator in order to make up the heat deficiency is most nearly

(A) 0.26 L/h

(B) 3.8 L/h

(C) 38 L/h

(D) 260 L/h

**52.** 1 mol of sodium hydroxide (NaOH) is added to 10 L of pure water at 25°C. The ion product for water at this temperature is $1 \times 10^{-14}$ mol²/L². The pH of the solution is most nearly

(A) 1

(B) 2

(C) 13

(D) 14

**53.** Federal underground storage tank regulations are primarily intended to protect

(A) workers at facilities with underground storage tanks

(B) groundwater quality

(C) air quality

(D) customers of facilities with underground storage tanks

**54.** The first-order reaeration constant of a stream depends on temperature according to the relationship

$$k(T) = \left(\frac{0.44}{h}\right)(1.02^{T-20°C})$$

At a wastewater outfall, the dissolved oxygen concentration in the stream is 5.8 mg/L. During the summer, when the stream's temperature is 30°C, the stream has a saturation dissolved oxygen concentration of 7.6 mg/L and a first-order reaeration coefficient of 0.54/h. Also in summer, the time that the stream takes to recover to a dissolved oxygen concentration of 7.3 mg/L after the outfall is 3.3 h. During the winter, the stream's temperature is 10°C and its saturation dissolved oxygen concentration is 11.3 mg/L. In the winter, the time the stream needs to go from a dissolved oxygen concentration of 5.8 mg/L to 7.3 mg/L is most nearly

(A) 3.8 times less than in summer

(B) the same as in summer

(C) 1.6 times more than in summer

(D) 3.8 times more than in summer

**55.** A source of x-rays strikes an individual who is 4 m away. There are no shields. If the individual moves to a point 8 m from the source, his or her exposure is reduced by a factor of

(A) 2
(B) 4
(C) 12
(D) 16

**56.** A well-mixed lake with a surface area of 3.0 km² and an average depth of 5 m is fed by a stream and by rainfall. (Groundwater and runoff are insignificant inputs.) The stream's flow rate is 0.2 m³/s, and the concentration of phosphorus in the stream is 0.1 mg/L. Average rainfall is 0.7 m/yr, and the rain has a phosphorus concentration of 0.01 mg/L. Phosphorus removal by evaporation and by biological processes is negligible. Phosphorus precipitates from the lake's water according to first-order kinetics, with a rate constant of 0.005/d. The outlet stream has a flow rate of 0.27 m³/s, which keeps the volume of water in the lake essentially constant. The concentration of phosphorus in the lake is most nearly

(A) 0.018 mg/L
(B) 0.029 mg/L
(C) 0.034 mg/L
(D) 0.045 mg/L

**57.** The overall nitrification reaction when ammonia is released to water is

$$NH_4^+ + 1.731 O_2 + 1.962 HCO_3^- \rightarrow 0.038 C_5H_7NO_2 + 0.962 NO_3^- + 1.077 H_2O + 1.769 H_2CO_3$$

In this equation, $C_5H_7NO_2$ represents bacterial cells. If this is the sole removal process when 30 kg of ammonia is released from a livestock operation to a stream, the amount of oxygen consumed is most nearly

(A) 2.9 kg $O_2$
(B) 52 kg $O_2$
(C) 76 kg $O_2$
(D) 92 kg $O_2$

**58.** Mixed wastes are wastes that

(A) contain radioactive wastes regulated under the Atomic Energy Act (AEA) and hazardous wastes regulated under the Resource Conservation and Recovery Act (RCRA)
(B) are comprised of both solids and liquids
(C) include both nonhazardous and hazardous wastes regulated under the RCRA
(D) contain municipal solid wastes and hazardous wastes regulated under the RCRA

**59.** Carbonates are the only significant source of inorganic carbon in a system. The presence of acid-neutralizing species other than carbonates is negligible. The pH of the system is 6.6 and the alkalinity is 160 mg/L as $CaCO_3$. The equilibrium constant for the first dissociation reaction of carbonate in this system is $10^{-6.4}$ mol/L, and the equilibrium constant for the second dissociation reaction of carbonate is $10^{-10.3}$ mol/L. Because carbonates are the only source of carbon, dissolved inorganic carbon (DIC) for this system is given by the following equation.

$$DIC = [H_2CO_3] + [HCO_3^-] + [CO_3^{2-}]$$

Alkalinity in terms of molar equivalents for this system is given by the following equation.

$$Alk = [HCO_3^-] + 2[CO_3^{2-}] + [OH^-] - [H^+]$$

The self-ionization constant for the dissociation of water in this system is $10^{-14}$ mol²/L². The DIC for this system is most nearly

(A) 31 mg C/L
(B) 63 mg C/L
(C) 67 mg C/L
(D) 120 mg C/L

**60.** The total amount of sulfur (in all forms) in the atmosphere is 4.0 Tg. Sulfur in $SO_4^{2-}$ is transferred to the oceans at a rate of 2.7 Mg/s. It is also transferred to the earth's crust at a rate of 3.2 Mg/s. If these are the only important removal processes for atmospheric sulfur, the steady-state residence time of sulfur is most nearly

(A) 0.13 d
(B) 7.8 d
(C) 14 d
(D) 17 d

**61.** For the equation shown, using Newton's method for root extraction and a starting point of $X_n = 1.1$ with $n = 0$, at which value does the convergence of the roots occur?

$$x(x^2 + 2) = 5$$

(A) $x = 1.2$

(B) $x = \sqrt{2}\,i$.

(C) $x = -\sqrt{2}\,i$

(D) $x = 1.32$

**62.** What is $x$ most nearly equal to at the minimum of the equation $f(x) = x^2 + 4x - 5$?

(A) $-2$

(B) 0

(C) 2

(D) 4

**63.** A particle moves along a straight line, with the displacement in meters being given by the empirical equation shown.

$$s = 2t^3 - t$$

$t$ is the time in seconds. When the acceleration of the particle is 24 m/s$^2$, the displacement is most nearly

(A) 0.5 m

(B) 10 m

(C) 12 m

(D) 14 m

**64.** The equation shown will form an ellipse when represented graphically on an $x$-$y$ plot.

$$\frac{(x-2)^2}{4} + \frac{(y-3)^2}{9} = 1$$

Where on the plot will the ellipse be located?

(A) center at (2, 3), major axis along $x = 2$, minor axis along $y = 3$

(B) center at (0, 0), major axis along $x = 2$, minor axis along $y = 3$

(C) center at $(-2, -3)$, major axis along $x = 2$, minor axis along $y = 3$

(D) center at (0, 0), major axis along the $y$ axis, minor axis along the $x$ axis

**65.** What is the solution to the differential equation shown?

$$\frac{d^2y}{dx^2} = 2xe^x + 3\cos x$$

(A) $2xe^x + 1$

(B) $2xe^x - 4e^x - 3\cos x + c_1 x + c_2$

(C) 1

(D) 0

**66.** Heat transfer occurs from a pipe to the surrounding environment due to natural convection processes. The flow rate in the pipe is 15 kg/min, and the area of the exposed pipe is 1.2 m$^2$. The average temperature of the fluid is 85°C. The temperature of the external environment is 20°C. The overall heat transfer coefficient is 7.50 W/m$^2$·K. The specific heat of the fluid is 0.95 J/g·°C. The expected change in the temperature after heat loss is most nearly

(A) 2.4°C

(B) 3.0°C

(C) 3.4°C

(D) 4.0°C

**67.** The dissolved oxygen (DO) at a point in a tank is measured over a 9 h period. The results are shown.

| time (h) | DO (mg/L) |
|---|---|
| 1 | 8.0 mg/L |
| 2 | 7.8 mg/L |
| 3 | 7.5 mg/L |
| 4 | 7.0 mg/L |
| 5 | 6.0 mg/L |
| 6 | 5.4 mg/L |
| 7 | 5.0 mg/L |
| 8 | 4.8 mg/L |
| 9 | 4.7 mg/L |

The mean DO during this period is

(A) 6.0 mg/L

(B) 6.2 mg/L

(C) 7.0 mg/L

(D) 7.5 mg/L

**68.** The dissolved oxygen (DO) at a point in a tank is measured over a 9 h period. The results are shown.

| time (h) | DO (mg/L) |
|---|---|
| 1 | 8.0 mg/L |
| 2 | 7.8 mg/L |
| 3 | 7.5 mg/L |
| 4 | 7.0 mg/L |
| 5 | 6.0 mg/L |
| 6 | 5.4 mg/L |
| 7 | 5.0 mg/L |
| 8 | 4.8 mg/L |
| 9 | 4.7 mg/L |

The median DO during this period is

(A) 5.0 mg/L

(B) 5.4 mg/L

(C) 6.0 mg/L

(D) 7.0 mg/L

**69.** Two tanks, each with a volume of 3.5 m$^3$, are connected by means of a small pipe with a valve. The tanks are made of rigid materials. Both tanks, the pipe, and the valve are completely insulated from the surroundings. One of the tanks contains a gas at a pressure of 1 atm and a temperature of 25°C, and the other tank is completely evacuated. The valve is initially closed. At time $t = 0$, the valve is opened. After a very long time, the temperature of the gas in both tanks is most nearly

(A) 0°C

(B) 12.5°C

(C) 20°C

(D) 25°C

**70.** A plate made of 4340 steel has a geometrical factor of 1. There is a 2 mm crack in the center of the plate. The applied engineering stress that will cause the plate to fail is most nearly

(A) 2.0 MPa

(B) 580 MPa

(C) 750 MPa

(D) 820 MPa

**71.** An ideal gas contains 40% nitrogen ($N_2$) by mass, 40% oxygen ($O_2$) by mass, and 20% carbon monoxide (CO) by mass. The gas is at a pressure of 2 atm. The partial pressure of nitrogen in the gas is most nearly

(A) 0.52 atm

(B) 0.64 atm

(C) 0.84 atm

(D) 1.04 atm

**72.** A fluid contains nitrogen ($N_2$) and oxygen ($O_2$) in a proportion of 80 g of $N_2$ to every 1 g of $O_2$. The fluid also has a 1% molar concentration of carbon dioxide ($CO_2$). The fluid is to be mixed with enough $N_2$ to reduce the molar concentration of $CO_2$ to 0.5. The total pressure in the fluid is 1.2 atm. The partial pressure of $N_2$ in the fluid after the extra $N_2$ is added is most nearly

(A) 1 atm

(B) 1.08 atm

(C) 1.15 atm

(D) 1.20 atm

**73.** A slab made up of two layers has a temperature difference of 100°C applied across it. Each layer is 1 m thick. The thermal conductivity of the outer layer is 1.5 kJ/m$^2$·K, and the thermal conductivity of the inner layer is 2.5 kJ/m$^2$·K. The amount of heat transmitted for one square meter of surface area is most nearly

(A) 48 kJ/m$^2$

(B) 58 kJ/m$^2$

(C) 68 kJ/m$^2$

(D) 78 kJ/m$^2$

**74.** If an engineering licensee's professional judgment is overruled by their employer under circumstances in which the health, safety, or welfare of the public is endangered, it is the licensee's obligation to the public to notify

(A) their employer and such other authorities as may be appropriate

(B) their employer only

(C) their licensing board

(D) the carrier of their professional liability insurance policy

**75.** An outstanding professional engineer (PE) in the electrical engineering discipline is also respected as an excellent project manager with good personal skills. They are offered a job as a manager at a consulting firm to supervise a group of PEs in civil, mechanical, electrical, and chemical engineering specializations. Which of these statements about the new manager's licensing is true?

(A) The engineer must obtain a license in a least one more discipline in order to take the job.

(B) The engineer could take the job without any additional licensure.

(C) The engineer needs to obtain licenses in all disciplines their supervised employees practice.

(D) The law forbids the engineer from taking the job.

**76.** The reaction rate for a given reaction is 10 times greater at 70°C than it is at 20°C. The activation energy of this reaction is most nearly

(A) 8.6 kcal/mol

(B) 9.0 kcal/mol

(C) 9.2 kcal/mol

(D) 9.7 kcal/mol

**77.** A forest-industries plant has a capacity utilization that is based on conditions in the forest. The possible percentages of capacity utilization for the next year, the probability each scenario will occur, and the net annual worth of the plant in that scenario are shown.

| capacity utilization | probability | net annual worth |
|---|---|---|
| 50% | 0.15 | $10,000,000 |
| 60% | 0.15 | $12,000,000 |
| 70% | 0.20 | $15,000,000 |
| 78% | 0.20 | $18,000,000 |
| 100% | 0.30 | $25,000,000 |

The expected value of the net annual worth of the plant is most nearly

(A) $10,000,000

(B) $12,000,000

(C) $17,000,000

(D) $20,000,000

**78.** A rounded venturi meter with a throat diameter of 1.5 in is placed in a 3.0 in horizontal pipe. The pressure drop is 3 mm Hg. The volumetric flow rate is most nearly

(A) 0.0009 m³/s

(B) 0.0022 m³/s

(C) 0.0033 m³/s

(D) 0.0044 m³/s

**79.** An open-door steady channel flow has a sudden decrease in the cross section for flow because of a decrease in width. Friction losses in the flow are negligible. After the decrease, the height of the liquid will

(A) increase

(B) decrease

(C) remain unchanged

(D) potentially change, depending on environmental pressure conditions

**80.** A legislative assembly committee needs to form a subcommittee consisting of two environmental experts and three lay persons, from a pool of six eligible environmental experts and six eligible lay persons. One specific environmental expert must be on the subcommittee. The number of ways the subcommittee can be formed is

(A) 100

(B) 200

(C) 300

(D) 400

**81.** A pitot tube is placed in a channel at a depth of 1 m. The stagnation head in the pitot tube is 1.6 m. The velocity of water at that point is most nearly

(A) 1.5 m/s

(B) 3.5 m/s

(C) 7.5 m/s

(D) 10.0 m/s

**82.** The cost associated with a remediation project is $10,000,000 if carried out now. There are three new environmental laws being considered for the state the project is located in, none of which are mutually exclusive, and two of which would increase the cost of the project if they are adopted. The costs under each of the new laws and the probabilities of each law being enacted are shown. All laws would take effect in one year. The prevailing interest rate in the state is 6%.

| law | probability | cost |
|---|---|---|
| 1 | 0.3 | $10,000,000 |
| 2 | 0.5 | $12,000,000 |
| 3 | 0.2 | $24,000,000 |

The expected change in the present worth of the costs if action is not taken until after one of the laws passes is most nearly

(A) 10%

(B) 18%

(C) 23%

(D) 30%

**83.** An environmental project gives benefits of $30,000,000 per year for 10 yr. The initial cost of the project is $10,000,000, and the final cost of the project is $5,000,000. The interest rate is 4%. The benefit-cost ratio is most nearly

(A) 2.0

(B) 12.0

(C) 18.0

(D) 20.0

**84.** An investment of $5,000,000 gives a $1,000,000 benefit annually for 10 yr. The rate of return on the investment is most nearly

(A) 5%

(B) 10%

(C) 15%

(D) 20%

**85.** A given radioactive compound decaying by a first-order reaction has a half-life of two years. The reaction rate constant is most nearly

(A) 0.35/yr

(B) 0.50/yr

(C) 0.75/yr

(D) 1.00/yr

**86.** Which of the following scenarios will occur if the indicated pairs of metals are immersed together in a solution that matches the conditions for measuring standard oxidation potentials?

(A) Aluminum will corrode if aluminum and iron are paired

(B) Cadmium will corrode if cadmium and zinc are paired

(C) There will be no corrosion if gold and palladium are paired

(D) There will be no corrosion if lead and copper are paired

**87.** The generic reaction for dehydration of alcohol is

$$+H^+ \rightleftharpoons +H_2O + H^+$$

The R' and/or R'' could be hydrogen. When 3-methyl-2-butanol is dehydrated, the major product of dehydration is most likely

(A) 3-methyl-1-butene

(B) 3-methyl-2-butene

(C) 2-methyl-2-butene

(D) 2-methyl-1-butene

**88.** An in situ bioassay using enclosed water columns is conducted to determine if algal growth in a lake during the summer months is limited by the concentration of nitrogen or phosphorus in the water. The concentration of chlorophyll A provides a relative measure of the growth of algal biomass in the lake. Results of the bioassay are given in the table.

| | chlorophyll A ($\mu g/L$) | | |
|---|---|---|---|
| time (d) | column #1: no additions (control) | column #2: addition of 70 $\mu g/L \cdot d$ nitrogen | column #3: addition of 35 $\mu g/L \cdot d$ phosphorus |
| 0 | 1.7 | 1.7 | 1.7 |
| 8 | 2.1 | 2.1 | 5 |

The results indicate that algal growth in the lake is likely to be limited by

(A) nitrogen
(B) phosphorus
(C) neither phosphorus nor nitrogen
(D) both nitrogen and phosphorus

**89.** Toxic gases can be generated if

(A) halogenated organics are combined with non-oxidizing mineral acids
(B) aromatic amines are combined with oxidizing mineral acids
(C) inorganic fluorides are combined with organic acids
(D) all of the above

**90.** The estimated amount of fish consumed by an adult in the U.S. over the course of a year is most nearly

(A) 2.2 kg
(B) 6 kg
(C) 8.9 kg
(D) 2200 kg

**91.** A study shows that the water supplied to homes in a particular neighborhood contains 23 ppb chloroform by mass. The dermal permeability constant for chloroform is 0.06 cm/hr. The daily absorbed dose of chloroform for a person in this neighborhood taking one shower of median length every day is most nearly

(A) 1.4 ng/kg·d
(B) 2.4 ng/kg·d
(C) 40 ng/kg·d
(D) 2400 ng/kg·d

**92.** A swimming pool has a chlorine concentration of 3 ppm. The chronic daily intake of chlorine due to ingesting pool water for a man swimming two times a week for half an hour at a time is most nearly

(A) 0.14 μg/kg·d
(B) 0.27 μg/kg·d
(C) 0.32 μg/kg·d
(D) 0.96 μg/kg·d

**93.** The runoff coefficient for a 20 ac tract of homes built on flat land (<2% slope) with average lot sizes of 0.25 ac is 0.45. For a rainfall event of 5 in/hr, the peak discharge from the tract is most nearly

(A) 0.23 ft³/sec
(B) 0.90 ft³/sec
(C) 11 ft³/sec
(D) 45 ft³/sec

**94.** For a given reservoir, area and active storage volume are related as shown.

$$A = \left(\frac{1.0}{\text{km}}\right)V + 7.0 \text{ km}^2$$

$A$ is the area of the reservoir in km², and $V$ is the reservoir's active storage volume in km³. For this reservoir, the average net loss from evaporation less rainfall during the summer is a function of area, as shown.

$$L = \left(6.4 \times 10^{-5} \frac{\text{km}}{\text{d}}\right)A$$

$L$ is the average net loss in km³/d. The active storage volume of the reservoir is 14 km³ during the summer. The additional inflow into the reservoir needed to make up for the average net loss over the course of the entire summer is most nearly

(A) 0.00091 km³
(B) 0.0013 km³
(C) 0.12 km³
(D) 0.47 km³

**95.** The equation for predicting average annual soil loss per unit area due to water erosion when crops and soil management are not involved is

$$A = RKL_S$$

$A$ is the average annual soil loss in short tons/ac, $R$ is the rainfall erosivity index, $K$ is the soil erodibility factor, and $L_S$ is the topographic factor. For a given empty lot that is 60 ft wide and 150 ft long, with a soil erodibility factor of 0.1, a topographic factor of 0.7, and a rainfall erosivity index of 350, the predicted average annual soil loss is most nearly

(A) 0.0027 short tons
(B) 5.1 short tons
(C) 10 short tons
(D) 25 short tons

**96.** The relationship for predicting the length of protection needed to prevent bend scour in a channel is shown.

$$\frac{L_P}{R_H} = 0.604\left(\frac{R_H^{1/6}}{n}\right)$$

$n$ is Manning's roughness coefficient, $R_H$ is the hydraulic radius, and $L_P$ is the recommended length of protection. A trapezoidal channel has a base of 4 m, and the sides of the channel have a slope of 25%. The depth of the water in the channel is 0.56 m. Manning's roughness coefficient for the channel walls is 0.03. The length of protection needed to prevent bend scour in this channel is most nearly

(A) 5.3 m
(B) 7.1 m
(C) 10 m
(D) 17 m

**97.** A water treatment flocculator has flat blades with a length-width ratio of 20:1 and a slip coefficient of 0.6. The area of the paddles is 30 m², and the paddle velocity is 0.5 m/s. The power input to the water being treated is most nearly

(A) 0.73 kW
(B) 1.5 kW
(C) 3.4 kW
(D) 8.1 kW

**98.** A test well is drilled in a confined well. The test well has a drawdown of 56 ft while pumping 9.4 gpm. The specific capacity of this well is most nearly

(A) 0.022 gpm/ft
(B) 0.17 gpm/ft
(C) 6.0 gpm/ft
(D) 45 gpm/ft

**99.** The seepage velocity of groundwater is

(A) the speed at which groundwater moves from point A to point B through the aquifer
(B) the volume of water passing through an entire unit cross section of aquifer per unit time
(C) the discharge rate of the aquifer
(D) equal to the hydraulic conductivity

**100.** Two identically constructed wells with the same pumping rate are placed in separate homogeneous confined aquifers. Pumping is begun at both wells at the same time. Aquifer 1 has a transmissivity of 1100 m²/d, and aquifer 2 has a transmissivity of 870 m²/d. At time $t$, the difference between the drawdown at distances of 5 m and 10 m from well 1 is 0.23 m. The difference between the drawdown at distances of 5 m and 10 m from well 2 at time $t$ is most nearly

(A) 0.18 m
(B) 0.23 m
(C) 0.29 m
(D) 330 m

**101.** The well shown reaches the bottom of a homogeneous confined aquifer, and has a radius of 0.15 m and a pumping rate of 32 L/s. The aquifer is 3.2 m deep. Hydraulic head at the well is 7.5 m, and hydraulic conductivity is 0.028 m/s.

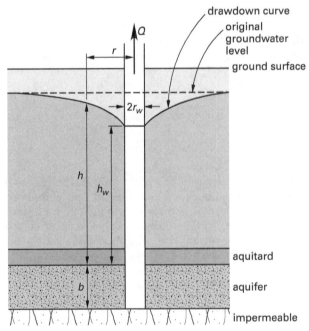

The hydraulic head 15 m away from the center of the well is most nearly

(A) 7.2 m
(B) 7.8 m
(C) 8.3 m
(D) 270 m

**102.** A confined, homogeneous aquifer has a storativity of $5 \times 10^{-4}$, a transmissivity of 370 ft²/day, and a pumping rate of 500 gpm. The Thiem equation for estimating drawdown in a confined, homogeneous aquifer is

$$s = \frac{QW(u)}{4\pi T}$$

$s$ is drawdown, $W(u)$ is the well function, $Q$ is the pumping rate, and $T$ is the aquifer transmissivity. The well function is

$$W(u) = -0.5772 - \ln u + u - \frac{u^2}{2(2!)} + \frac{u^3}{3(3!)} - \frac{u^4}{4(4!)} + \cdots$$

$u$ is the dimensionless well function argument, which is given by the equation shown.

$$u = \frac{r^2 S}{4Tt}$$

In this equation, $r$ is the distance from the well, $S$ is the storativity of the aquifer, $T$ is the aquifer transmissivity, and $t$ is the time that has elapsed since pumping began. The drawdown at a distance of 1000 ft from a well in the confined, homogeneous aquifer 338 days after pumping begins is most nearly

(A) 0.021 ft

(B) 0.068 ft

(C) 130 ft

(D) 270 ft

**103.** The Darcy velocity of a non-aqueous phase liquid (NAPL) in soil that is saturated by the NAPL can be found using the equation shown.

$$v = \frac{k\rho g}{\mu}\left(\frac{dh}{dx}\right)$$

Two contaminants are released to identical parcels of soil such that their hydraulic gradients are the same. Contaminant A has a density of 1 g/cm³ and a dynamic viscosity of 0.001 Pa/s. Contaminant B has a density of 0.89 g/cm³ and a dynamic viscosity of 0.004 Pa/s. The speed at which contaminant A migrates through the soil is most nearly

(A) 0.22 times that of contaminant B

(B) 0.28 times that of contaminant B

(C) 3.6 times that of contaminant B

(D) 4.5 times that of contaminant B

**104.** Unsaturated soil at 20°C is contaminated with benzene. Remediation will occur via soil vapor extraction with the benzene adsorbed onto activated carbon. The desired concentration of benzene after adsorption is 0.02 g/m³. The constants $K$ and $n$ for the Freundlich isotherm are 0.012 and 1.9, respectively, for an equilibrium concentration in units of g/m³ and a mass ratio in units of g/g. The air flow from the well is 0.3 m³/min, and the concentration of benzene in the pumped air is constant at 10 mg/L. The volume of fresh activated carbon needed each day is most nearly

(A) 4.3 kg/d

(B) 2000 kg/d

(C) 2800 kg/d

(D) 600 000 kg/d

**105.** A fresh organic material contains 40% water and contains 0.18 kg carbon/kg dry solids. A minimum of 65% of the carbon in the fresh organic material is converted to carbon dioxide during composting. The minimum annual carbon dioxide emissions rate for a facility that composts 6000 Mg/y of the fresh organic material is most nearly

(A) 420 Mg/y

(B) 1000 Mg/y

(C) 1500 Mg/y

(D) 3900 Mg/y

**106.** Two municipal waste streams will be composted. The first stream is spoiled produce with a carbon-nitrogen mass ratio of 18. The second stream is cardboard with a carbon-nitrogen ratio of 400. The ideal carbon-nitrogen mass ratio for composting is 30. The ideal percentage by mass of the mixed streams that is spoiled produce is most nearly

(A) 91%

(B) 93%

(C) 96%

(D) 97%

**107.** Insufficient nitrogen in a composting process is a problem because

(A) the composting process slows when nitrogen is insufficient

(B) a compost pile with insufficient nitrogen is likely to undergo spontaneous combustion

(C) odor problems are associated with insufficient nitrogen

(D) a compost pile with insufficient nitrogen is likely to become too dry

**108.** A machine shop produces 1200 aluminum parts in one year, each of which weighs 0.5 kg. During that year, the shop purchases 1000 kg of aluminum and generates 100 kg of chips and scrap that are shipped to recycling. At the beginning of the year the aluminum inventory is 1000 kg, and at the end of the year the aluminum inventory is 1100 kg. The only other significant stream of aluminum is aluminum dust, which is landfilled. The amount of aluminum landfilled during the year is most nearly

(A) 100 kg
(B) 200 kg
(C) 300 kg
(D) 400 kg

**109.** A hazardous waste containing inorganic fluorides cannot safely be mixed with

(A) glycols
(B) caustics
(C) cyanides
(D) organic acids

**110.** The total residential consumption of natural gas in the US in 2016 was 120 000 000 m³. Assume natural gas is 100% methane, and that all the carbon in the natural gas is converted to carbon dioxide during combustion. The carbon dioxide emitted due to residential natural gas combustion in 2016 was most nearly

(A) 440 Mg
(B) 20 000 Mg
(C) 64 000 Mg
(D) 240 000 Mg

## SOLUTIONS

**1.** A waste sample is considered corrosive if it is aqueous and has a pH of either less than 2 or greater than 12.5.

**The answer is (B).**

**2.** Lime is found to be useful in accelerating the sedimentation of sewage.

**The answer is (B).**

**3.** From a water properties table, the density of water at 25°C is 997 kg/m³. [**Properties of Water (SI Metric Units)**]

It is given that the average velocity of flow is 0.15 m/s. The cross-sectional area of the channel is

$$A = (4 \text{ m})(1 \text{ m}) + \left(\frac{1}{2}\right)(2)(2 \text{ m})(4 \text{ m})$$
$$= 12 \text{ m}^2$$

Therefore, the mass flow rate is

$$\dot{m} = \rho v A$$
$$= \left(997 \; \frac{\text{kg}}{\text{m}^3}\right)\left(0.15 \; \frac{\text{m}}{\text{s}}\right)(12 \text{ m}^2)$$
$$= 1800 \text{ kg/s}$$

**The answer is (B).**

**4.** The difference between the atmospheric pressure and the pressure in the tank is

$$p_t - p_a = (\rho_{\text{man}} - \rho_{H_2O})gh$$

In this equation, $\rho_{\text{man}}$ is the density of the manometer's fluid and is equal to $5.6\rho_{H_2O}$, and $\rho_{H_2O}$ is the density of water in the tank (at 20°C) and is equal to 998.2 kg/m³. Substituting values gives

$$p_t - p_a = \left((5.6 - 1.0)\left(998.2 \; \frac{\text{kg}}{\text{m}^3}\right)\right)$$
$$\times \left(9.81 \; \frac{\text{m}}{\text{s}^2}\right)(10 \text{ cm})\left(\frac{1 \text{ m}}{100 \text{ cm}}\right)$$
$$= 4504 \text{ kg·m/m}^2\text{·s}^2 \quad (4504 \text{ Pa})$$

The barometric pressure is

$$p_a = (750 \text{ mm Hg})\left(\frac{1.013 \times 10^5 \text{ Pa}}{760 \text{ mm Hg}}\right) = 99\,967 \text{ Pa}$$

Solving for $p_t$ and substituting gives

$$p_t = p_a + 4504 \text{ Pa}$$
$$= 99\,967 \text{ Pa} + 4504 \text{ Pa}$$
$$= 104\,000 \text{ Pa} \quad (100 \text{ kPa})$$

**The answer is (C).**

**5.** Let $A$ represent the material of interest. Assume that the fully mixed reactor can be modeled as a continuous stirred tank reactor (CSTR). It is given that the volume of the mixed tank, $V_{CSTR}$, is 5 L, the inflow concentration, $C_{A0}$, is 150 mg/L, and the conversion in the reactor, $X_A$, is 90%. Use the formula for a CSTR.

**Batch Reactor, Variable Volume**

$$\frac{V_{CSTR}}{F_{A0}} = \frac{X_A}{-r_A}$$

The moles of $A$ fed per unit time is

$$F_{A0} = QC_{A0}$$

Also, the concentration of $A$ leaving the reactor is

$$C_A = C_{A0}(1 - X_A)$$
$$= \left(150 \, \frac{\text{mg}}{\text{L}}\right)(1 - 0.9)$$
$$= 15 \text{ mg/L}$$

The reaction rate in the tank, $r_A$, is 50/d. From the units of this rate constant, it can be deduced that the reaction is a first-order reaction. Therefore, the rate can be expressed as

$$-r_A = kC_A$$
$$= \left(50 \, \frac{1}{\text{d}}\right)\left(15 \, \frac{\text{mg}}{\text{L}}\right)$$
$$= 750 \text{ mg/L·d}$$

Substituting and solving for $Q$ gives

$$Q = -\frac{r_A V_{CSTR}}{C_{A0} X_A}$$
$$= \frac{\left(750 \, \frac{\text{mg}}{\text{L·d}}\right)(5 \text{ L})\left(\frac{1 \text{ d}}{24 \text{ h}}\right)\left(\frac{1 \text{ h}}{60 \text{ min}}\right)}{\left(150 \, \frac{\text{mg}}{\text{L}}\right)\left(\frac{1 \text{ L}}{1000 \text{ cm}^3}\right)(0.9)}$$
$$= 19 \text{ cm}^3/\text{min}$$

**The answer is (A).**

**6.** The coefficient of the meter, $C$, is 0.61 (from a table of orifices and their nominal coefficients). Assuming that the pipe is horizontal, the volumetric flow rate is

$$Q = CA\sqrt{\frac{2}{\rho}(p_1 - p_2)}$$

The cross-sectional area at the orifice is

$$A = \frac{\pi d^2}{4}$$

The density of water at 25°C is 997 kg/m³, and the pressure drop, $p_1 - p_2$, is given as 10 Pa. [**Properties of Water (SI Metric Units)**]

Therefore,

$$Q = (0.61)\left(\frac{\pi(0.046 \text{ m})^2}{4}\right)$$
$$\times \sqrt{\left(\frac{2}{997 \, \frac{\text{kg}}{\text{m}^3}}\right)(10 \text{ Pa})\left(1 \, \frac{\frac{\text{N}}{\text{m}^2}}{\text{Pa}}\right)} \times \left(1 \, \frac{\frac{\text{kg·m}}{\text{s}^2}}{\text{N}}\right)\left(3600 \, \frac{\text{s}}{\text{h}}\right)^2$$

$$= 0.51 \text{ m}^3/\text{h}$$

**The answer is (D).**

**7.** The average velocity at any value of $d$ is given at the vertical midpoint, or $0.5h$. Since the given velocities in the problem statement are for $0.3h$ and $0.7h$ for each distance $d$, the velocity at $0.5h$ is interpolated using the formula

$$v_{0.5h} = \frac{v_{0.3h} + v_{0.7h}}{2}$$

The interpolated values are given in the following table.

| distance from the left bank, $d$ (m) | depth of river, $h$ (m) | velocity of flow, v (m/s) | | |
|---|---|---|---|---|
| | | $0.3h$ | $0.7h$ | $0.5h$ |
| 1 | 1 | 0.5 | 0.5 | 0.50 |
| 2 | 3 | 0.6 | 0.3 | 0.45 |
| 3 | 4 | 0.8 | 0.6 | 0.70 |
| 4 | 5 | 0.9 | 0.6 | 0.75 |
| 5 | 5 | 0.9 | 0.6 | 0.75 |
| 6 | 5 | 0.9 | 0.6 | 0.75 |
| 7 | 5 | 1.0 | 0.8 | 0.90 |
| 8 | 2.5 | 0.6 | 0.4 | 0.50 |

The sum of the velocities at $0.5h$ is 5.3 m/s. Therefore, average velocity is

$$\text{average velocity} = \frac{\text{sum of velocities}}{\text{number of data points}}$$

$$= \frac{5.3 \; \frac{\text{m}}{\text{s}}}{8}$$

$$= 0.66 \text{ m/s}$$

**The answer is (C).**

**8.** For a first-order reaction, the rate constant when the half-life is known can be calculated from the equation shown.

**Half-Life**

$$k = \frac{0.693}{t_{1/2}}$$

Solving for $t_{1/2}$ gives

$$t_{1/2} = \frac{0.693}{k}$$

$$= \frac{0.693}{0.067 \; \frac{1}{\text{h}}}$$

$$= 10 \text{ h}$$

**The answer is (B).**

**9.** Let $t$ be the time elapsed (in years) since 1980. The per-capita water consumption can be taken to be a constant of 47 L/d/person. At full capacity, $C$, the pumping system can deliver 76 million L/d. This value is equal to the population at that time multiplied by 47 L/person/d. In equation form, this is given by

$$C = P\left(47 \; \frac{\text{L}}{\text{d}}\right) = 76 \times 10^6 \text{ L/d}$$

Solving for $P$ gives

$$P = \frac{76 \times 10^6 \; \frac{\text{L}}{\text{d}}}{47 \; \frac{\text{L}}{\text{person} \cdot \text{d}}} = 1.6 \text{ million people}$$

The curve-fit formula for the population as a function of time can be given as

$$P = 1.1 \times 10^6 \text{ people} + \left(\frac{0.1 \times 10^6 \text{ people}}{10 \text{ yr}}\right)t$$

$$= 1.1 \times 10^6 \text{ people} + \left(10\,000 \; \frac{\text{people}}{\text{yr}}\right)t$$

Substitute values and solve for $t$.

$$t = \frac{1.6 \times 10^6 \text{ people} - 1.1 \times 10^6 \text{ people}}{10\,000 \; \frac{\text{people}}{\text{yr}}}$$

$$= 50 \text{ yr}$$

Therefore, the year during which there will be a need for more capacity is

$$1980 + 50 = 2030$$

**The answer is (C).**

**10.** The population, $P$, that would be reached when the pumping station capacity, $C$, is fully utilized would be

$$P = \frac{C}{36 \; \frac{\text{L}}{\text{person} \cdot \text{d}}}$$

$$= \frac{110 \times 10^6 \; \frac{\text{L}}{\text{d}}}{36 \; \frac{\text{L}}{\text{person} \cdot \text{d}}}$$

$$= 3.1 \times 10^6 \text{ people}$$

Since the population does not appear to be growing linearly, an exponential formula of the following form can be tried.

$$P = P_0 e^{kt}$$

Solving for $k$ gives

$$k = \frac{\ln P - \ln P_0}{t}$$

If ln P changes linearly with time, then the data fit an exponential curve. In tabular form, the population data are as follows.

| year, relative to present (yr) | $\Delta t$ (yr) | P, million persons | ln P |
|---|---|---|---|
| −20 | 10 | 1.2 | 14.0 |
| −10 | 10 | 1.3 | 14.1 |
| 0 | 10 | 1.5 | 14.2 |

This table shows that the population data follow an exponential curve.

$$k = \frac{14.1 - 14.0}{10 \text{ yr}} = \frac{14.2 - 14.0}{20 \text{ yr}} = \frac{14.2 - 14.1}{10 \text{ yr}}$$

$$= \frac{0.1}{10 \text{ yr}}$$

$$= 0.01/\text{yr}$$

The population curve is

$$P = (1.5 \times 10^6 \text{ people}) e^{\left(0.01 \frac{1}{\text{yr}}\right) t}$$

Solve for $t$ to get

$$t = \frac{\ln P - \ln 1.5 \times 10^6 \text{ people}}{0.01 \frac{1}{\text{yr}}}$$

$$= \frac{\ln 3.1 \times 10^6 \text{ people} - \ln 1.5 \times 10^6 \text{ people}}{0.01 \frac{1}{\text{yr}}}$$

$$= 73 \text{ yr}$$

Full capacity will be utilized 73 yr in the future.

**The answer is (C).**

**11.** Set the two equations equal to each other.

$$S = \frac{\eta \gamma_w b g}{E_w B} = \beta \gamma_w b g$$

Solve for $B$, and substitute.

$$B = \frac{\eta}{E_w \beta}$$

$$= \frac{0.2}{\left(2.07 \times 10^9 \frac{\text{N}}{\text{m}^2}\right)\left(1.9 \times 10^{-7} \frac{\text{m}^2}{\text{N}}\right)}$$

$$= 0.00051$$

**The answer is (B).**

**12.** First, find the concentration of hydrogen ions in the solution.

**Acids, Bases, and pH**

$$\text{pH} = \log_{10}\left(\frac{1}{[H^+]}\right)$$

$$= \frac{1}{10^2}$$

$$= 0.01 \text{ mol/L}$$

NaOH donates one hydroxide ion per molecule and has a molecular weight of 40 g/mol, so the mass of NaOH required is

$$m_{\text{NaOH}} = V[H^+] \text{MW}_{\text{NaOH}}$$

$$= (1000 \text{ L})\left(0.01 \frac{\text{mol}}{\text{L}}\right)\left(40 \frac{\text{g NaOH}}{\text{mol}}\right)$$

$$= 400 \text{ g NaOH}$$

**The answer is (C).**

**13.** The diameter of pipe, $D$, is 2 cm (0.02 m). At 27°C, the density of water, $\rho$, is 996.0 kg/m³, the viscosity of water, $\mu$, is $0.854 \times 10^{-3}$ Pa·s, and the kinematic viscosity of water, $\mu/\rho$, is $0.857 \times 10^{-6}$ m²/s. The volumetric flow rate, $Q$, is $1 \times 10^{-6}$ m³/s. The pressure drop is obtained using the Darcy equation.

**Head Loss Due to Flow**

$$\frac{-\Delta p}{\gamma} = f \frac{L}{D} \frac{\text{v}^2}{2g}$$

The velocity of flow is given by

$$\text{v} = \frac{Q}{A} = \frac{Q}{\frac{\pi D^2}{4}}$$

$$= \frac{1 \times 10^{-6} \frac{\text{m}^3}{\text{s}}}{\frac{\pi (0.02 \text{ m})^2}{4}}$$

$$= 3.2 \times 10^{-3} \text{ m/s}$$

The friction factor can be found from the Moody diagram, after computing the Reynolds number, Re. [**Flow in Closed Conduits**]

### Transport Phemonena-Momentum, Heat, and Mass-Transfer Analogy

$$Re = \frac{Dv\rho}{\mu}$$

$$= \frac{(0.02 \text{ m})\left(3.2 \times 10^{-3} \frac{\text{m}}{\text{s}}\right)\left(996 \frac{\text{kg}}{\text{m}^3}\right)}{0.854 \times 10^{-3} \text{ Pa·s}}$$

$$= 75 \quad [\text{laminar flow}]$$

For laminar flow, the friction factor is

$$f = \frac{64}{Re} = \frac{64}{75} = 0.85$$

Remember that

$$\gamma = \rho g$$

This means that the Darcy equation can be written as

$$-\Delta p = \frac{\rho f L v^2}{2D}$$

$$= \frac{\left(996 \frac{\text{kg}}{\text{m}^3}\right)(0.85)(1 \text{ m})\left(3.2 \times 10^{-3} \frac{\text{m}}{\text{s}}\right)^2}{(2)(0.02 \text{ m})}$$

$$= -0.22 \text{ Pa}$$

This is the change in pressure for 1 m of pipe, so there is a pressure drop of 0.22 Pa/m of pipe.

**The answer is (B).**

**14.** The retardation factor, $R$, is the average groundwater velocity divided by the velocity of the adsorbed material, or

$$v_c = \frac{v_g}{R}$$

Solve for $v_g$, and substitute.

$$v_g = v_c R$$
$$= 5v_c$$

Therefore, the groundwater velocity is five times the contaminant velocity.

**The answer is (C).**

**15.** Initial volume is 18.5 m³. Reduction in volume is 85%.

final volume = (1 − fractional reduction in volume)
 × (initial volume)
 = (1 − 0.85)(18.5 m³)
 = 2.8 m³

$$\text{compaction ratio} = \frac{\text{initial volume}}{\text{final volume}}$$

$$= \frac{18.5 \text{ m}^3}{2.8 \text{ m}^3}$$

$$= 6.7$$

**The answer is (B).**

**16.** Bernoulli's equation is applied between the top of the tank and the exit point of fluid from the pipe.

$$\frac{p_1}{\rho} + gz_1 + \frac{v_1^2}{2} + \eta W_p = \frac{p_2}{\rho} + gz_2 + \frac{v_2^2}{2} + h_f g$$

In this equation, $p_1$ is the pressure at the top of the tank (1 atm), $p_2$ is the pressure at the exit point of the pipe (1 atm), $\rho$ is the density of water at 20°C (998.2 kg/m³), $g$ is the acceleration of gravity (9.81 m/s²), $z_1$ is the height of the top of the tank above the datum (20 m), $z_2$ is the height of the top of the adsorption system above the datum (100 m), $v_1$ is the velocity of the fluid at the top of tank (0 m/s, assuming that the tank has a large cross-sectional area), $\eta$ is the efficiency of the pump (0.60), $h_f$ is the head loss in the piping system (1 m), and $W_p$ is the energy consumption in the pump per unit mass of fluid.

Bernoulli's equation simplifies to

$$\eta W_p = g(z_2 - z_1) + \frac{v_2^2}{2} + h_f$$

The velocity of flow in the pipe is

$$v_2 = \frac{Q}{A} = \frac{Q}{\frac{\pi D^2}{4}}$$

The diameter of the pipe, $D$, is 4 cm (0.04 m), and the volumetric flow rate is 1 L/s. Substituting and solving for $W_p$ gives

$$\eta W_p = g(z_2 - z_1) + \frac{\left(\dfrac{Q}{\dfrac{\pi D^2}{4}}\right)^2}{2} + h_f g$$

$$0.60\, W_p = \left(9.81\ \frac{\text{m}}{\text{s}^2}\right)(100\ \text{m} - 20\ \text{m})$$

$$+ \frac{\left(\dfrac{\left(1\ \dfrac{\text{L}}{\text{s}}\right)\left(\dfrac{1\ \text{m}^3}{1000\ \text{L}}\right)}{\dfrac{\pi(0.04\ \text{m})^2}{4}}\right)^2}{2}$$

$$+ (1\ \text{m})\left(9.81\ \frac{\text{m}}{\text{s}^2}\right)$$

$$W_p = 1300\ \text{m}^2/\text{s}^2$$

The power of the pump is

$$P = W_p Q \rho$$
$$= \left(1300\ \frac{\text{m}^2}{\text{s}^2}\right)\left(1\ \frac{\text{L}}{\text{s}}\right)\left(\frac{1\ \text{m}^3}{1000\ \text{L}}\right)\left(997.2\ \frac{\text{kg}}{\text{m}^3}\right)$$
$$= 1.3 \times 10^3\ \text{kg·m·m/s}^2\text{·s} \quad (1.3\ \text{kW})$$

**The answer is (C).**

**17.** A discharge wetland is one where the wetland's water surface is below the water table of the surrounding land. Groundwater is discharged into such a wetland. Option A describes a recharge wetland, option C describes a perched wetland, and option D describes a spring or seep.

**The answer is (B).**

**18.** The pressure at the outlet of the pump can be obtained by applying Bernoulli's equation to flow through the pump. Point 1 is the inlet to the pump, and point 2 is the outlet of the pump.

$$\frac{p_1}{\gamma} + \frac{\text{v}_1^2}{2g} + z_1 + \frac{\eta W_p}{g} = \frac{p_2}{\gamma} + \frac{\text{v}_2^2}{2g} + z_2 + h_{f,\text{pump}}$$

It can be assumed that the velocities at the inlet and outlet parts are equal (i.e., $\text{v}_1 = \text{v}_2$) and that the heights of the inlet and outlet parts are approximately equal (i.e., $z_1 = z_2$). The head loss due to friction, $h_{f,\text{pump}}$, can be taken to be approximately zero, since its effect is included in the value of efficiency, $\eta$. Thus, Bernoulli's equation simplifies to

$$\frac{p_1}{\gamma} + \frac{\eta W_p}{g} = \frac{p_2}{\gamma}$$

Remember that

$$\gamma = \rho g$$

Substituting and solving for $p_2$ gives

$$p_2 = p_1 + \eta \rho W_p$$

It is given that $p_1$ is 69 kPa, that the pump power is 11 kW, and that the pump efficiency is 0.48. From a water properties table, the density of water at 25°C is 997 kg/m³. [**Properties of Water (SI Metric Units)**]

Find the value of $W_p$.

$$W_p = \frac{\text{energy consumed per unit time}}{\text{mass pumped per unit time}}$$
$$= \frac{\text{pump power}}{Q\rho}$$
$$= \frac{11\ \text{kW}}{\left(0.1\ \dfrac{\text{L}}{\text{s}}\right)\left(\dfrac{1\ \text{m}^3}{1000\ \text{L}}\right)\left(997\ \dfrac{\text{kg}}{\text{m}^3}\right)}$$
$$= 110\ \text{kJ/kg} \quad (110\,000\ \text{J/kg})$$

Substituting values yields

$$p_2 = 69 \times 10^3\ \text{Pa} + (0.48)\left(997\ \frac{\text{kg}}{\text{m}^3}\right)\left(110\,000\ \frac{\text{J}}{\text{kg}}\right)$$
$$= 5.3 \times 10^7\ \text{Pa}$$

**The answer is (D).**

**19.** The head loss in each pipe (from the Darcy equation) is

**Head Loss Due to Flow**

$$h_f = f\frac{L}{D}\frac{\text{v}^2}{2g}$$

Velocity in the first pipe, which has a flow rate of 10 L/min and a diameter of 4 cm (0.04 m), is

$$v = \frac{Q}{A} = \frac{Q}{\frac{\pi D^2}{4}}$$

$$= \frac{\left(10 \ \frac{L}{min}\right)\left(\frac{1 \ m^3}{1000 \ L}\right)\left(\frac{1 \ min}{60 \ s}\right)}{\frac{\pi(0.04 \ m)^2}{4}}$$

$$= 0.13 \ m/s$$

The friction factor, $f$, is a function of the Reynolds number. To use the graph of friction factor vs. Reynolds number in the *NCEES Handbook*, first calculate the Reynolds number.

$$Re = \frac{DV\rho}{\mu}$$

$$= \frac{(0.04 \ m)\left(0.13 \ \frac{m}{s}\right)\left(998.2 \ \frac{kg}{m^3}\right)}{1.002 \times 10^{-3} \ Pa \cdot s}$$

$$= 5200$$

Then calculate relative roughness. The roughness of the cast-iron pipe, $e$, is 0.25 mm (0.25 × 10⁻³ m). [**Flow in Closed Conduits**]

Therefore, the roughness factor is

$$\frac{e}{D} = \frac{0.25 \times 10^{-3} \ m}{0.04 \ m} = 0.00625$$

From friction factor chart, $f$ is 0.044. [**Flow in Closed Conduits**]

This means that head loss in the first pipe is

$$h_1 = \frac{(0.044)(10 \ m)\left(0.13 \ \frac{m}{s}\right)^2}{(0.04 \ m)(2)\left(9.81 \ \frac{m}{s^2}\right)}$$

$$= 0.0095 \ m$$

The head loss in the second pipe can be found in a similar manner. This pipe has a 2 cm (0.02 m) diameter and a flow velocity of

$$v = \frac{Q}{A} = \frac{Q}{\frac{\pi D^2}{4}}$$

$$= \frac{\left(10 \ \frac{L}{min}\right)\left(\frac{1 \ m^3}{1000 \ L}\right)\left(\frac{1 \ min}{60 \ s}\right)}{\frac{\pi(0.02 \ m)^2}{4}}$$

$$= 0.53 \ m/s$$

Calculate the Reynolds number.

$$Re = \frac{DV\rho}{\mu}$$

$$= \frac{(0.02 \ m)\left(0.53 \ \frac{m}{s}\right)\left(998.2 \ \frac{kg}{m^3}\right)}{1.002 \times 10^{-3} \ Pa \cdot s}$$

$$= 11\,000$$

The roughness factor is

$$\frac{e}{D} = \frac{0.25 \times 10^{-3} \ m}{0.02 \ m} = 0.0125$$

From a friction factor chart, $f$ is 0.045. [**Flow in Closed Conduits**]

This means that head loss in the second pipe is

$$h_2 = \frac{(0.045)(10 \ m)\left(0.53 \ \frac{m}{s}\right)^2}{(0.02 \ m)(2)\left(9.81 \ \frac{m}{s^2}\right)}$$

$$= 0.32 \ m$$

The head loss at the expansion between the two pipes can be taken as the nominal value for head loss in well-streamlined gradual contractions.

**Minor Losses in Pipe Fittings, Contractions, and Expansions**

$$h_{f,\text{fitting}} = \frac{0.04 v_{4\,cm}^2}{2g}$$

$$= \frac{(0.04)\left(0.13 \ \frac{m}{s}\right)^2}{(2)\left(9.81 \ \frac{m}{s^2}\right)}$$

$$= 3.4 \times 10^{-5} \ m$$

The total head loss is

$$h_{total} = h_1 + h_2 + h_{f,fitting}$$
$$= 0.0095 \text{ m} + 0.32 \text{ m} + 3.4 \times 10^{-5} \text{ m}$$
$$= 0.33 \text{ m}$$

The equivalent length, Le, for a 4 cm pipe can be found from the Darcy equation.

$$h_{total} = \frac{f_{4\text{ cm}} v_{4\text{ cm}}^2 \text{Le}}{2gD_{4\text{ cm}}}$$

From above, $f_{4\text{ cm}}$ is 0.044 and $v_{4\text{ cm}}$ is 0.13 m/s. Substituting and solving for Le gives

$$\text{Le} = \frac{2gD_{4\text{ cm}} h_{total}}{f_{4\text{ cm}} v_{4\text{ cm}}^2}$$
$$= \frac{(2)\left(9.81 \frac{\text{m}}{\text{s}^2}\right)(0.04 \text{ m})(0.33 \text{ m})}{(0.044)\left(0.13 \frac{\text{m}}{\text{s}}\right)^2}$$
$$= 350 \text{ m}$$

**The answer is (D).**

**20.** Kiln rake and rotational speed are inversely proportional to residence time. Halving either of these while leaving the other parameters constant doubles the residence time. [**Kiln Formula**]

**The answer is (B).**

**21.** Since the stream is 10% water vapor, it is 90% organic compounds. Therefore,

$$E_{TOC} = F_A \text{WF}_{TOC} N$$
$$= \left(0.0268 \frac{\frac{\text{kg}}{\text{h}}}{\text{valve}}\right)(0.9)(110 \text{ valves})$$
$$\times \left(350 \frac{\text{d}}{\text{yr}}\right)\left(24 \frac{\text{h}}{\text{d}}\right)$$
$$= 22\,300 \text{ kg/yr}$$

**The answer is (C).**

**22.** The ideal gas law is

$$PV = nRT$$

Solve for $n$.

$$n = \frac{PV}{RT}$$
$$= \frac{(5 \times 10^{-9} \text{ atm})(1 \text{ m}^3)\left(1000 \frac{\text{L}}{\text{m}^3}\right)}{\left(0.08206 \frac{\text{L} \cdot \text{atm}}{\text{mol} \cdot \text{K}}\right)(293\text{K})}$$
$$= 2 \times 10^{-7} \text{ mol}$$

**The answer is (B).**

**23.** For an ideal gas, parts per million (volume) is equivalent to liters per million liters of air. There are 22.4 L of carbon dioxide in each mole at standard temperature and pressure. The equation is

$$x \text{ ppm CO}_2 \text{ (volume)} = \left(\frac{x \text{ L CO}_2}{10^6 \text{ L air}}\right)\left(\frac{1 \text{ mol CO}_2}{22.4 \text{ L CO}_2}\right)$$

Therefore,

$$350 \text{ ppm CO}_2 \text{ (volume)} = \left(\frac{350 \text{ L CO}_2}{10^6 \text{ L air}}\right)\left(\frac{1 \text{ mol CO}_2}{22.4 \text{ L CO}_2}\right)$$
$$= 1.6 \times 10^{-5} \text{ mol/L}$$

**The answer is (B).**

**24.** The dimensionless parameter specific speed is defined as

$$n_s = \frac{n\sqrt{Q}}{h^{3/4}}$$

In this equation, the flow rate, $Q$, must be in gal/min, the head, $h$, must be in ft, and the rotational speed, $n$, must be in rpm. Substituting values yields

$$n_s = \frac{1000 \frac{\text{rev}}{\text{min}} \sqrt{10 \frac{\text{gal}}{\text{min}}}}{(100 \text{ ft})^{3/4}}$$
$$= 100$$

**The answer is (A).**

**25.** Find the mass of carbon in a mole of acetone.

$$\text{mass of carbon} = (3)\left(12 \frac{\text{g}}{\text{mol}}\right)$$
$$= 36 \text{ g carbon/mol acetone}$$

Find the molecular weight of acetone.

$$MW_{acetone} = (3)\left(12\ \frac{g}{mol}\right) + (6)\left(1\ \frac{g}{mol}\right) + (1)\left(16\ \frac{g}{mol}\right)$$
$$= 58\ g/mol$$

$$TOC = \frac{36\ g\ carbon}{58\ g\ acetone}$$
$$= 0.62\ g\ carbon/g\ acetone$$

**The answer is (D).**

**26.** The relationship between the head generated, $h$, and impeller diameter, $D$, for a centrifugal pump at constant rotational speed (from the scaling law formula) is

**Fans, Pumps, and Compressors**

$$\frac{h_2}{h_1} = \left(\frac{D_2}{D_1}\right)^2$$

Solving for $D_2$ and substituting the given values,

$$D_2 = D_1\sqrt{\frac{h_2}{h_1}}$$
$$= 1\ m\sqrt{\frac{2000\ m}{100\ m}}$$
$$= 4.5\ m$$

**The answer is (A).**

**27.** The first-order chemical reaction can be written as

$$\frac{-dL}{dt} = kL$$

In this equation, $k$ is the rate constant and $L$ is the BOD remaining at time $t$ (from the basic definition of BOD). Therefore,

$$\frac{-dL}{L} = kdt$$

Integrating from $t = 0$ d to $t = 5$ d and $L = L_0$ to $L = L_5$ gives

$$L_5 = L_0 e^{-k(5\ d)}$$

The 5-day BOD for the diluted sample, $y_5$, becomes

$$y_5 = L_0 - L_5 = L_0(1 - e^{-k(5\ d)})$$

The 5-day BOD is given as 5 mg/L, and the rate constant is given as 0.1/d. Solving for $L_0$ and substituting gives

$$L_0 = \frac{y_5}{1 - e^{-k(5\ d)}}$$
$$= \frac{5\ \frac{mg}{L}}{1 - e^{-\left(0.1\ \frac{1}{d}\right)(5\ d)}}$$
$$= 13\ mg/L$$

Since the ultimate BOD of the diluted sample is 13 mg/L, the ultimate BOD of the original sample is

$$\text{ultimate BOD} = (\text{ultimate BOD of diluted sample})$$
$$\times (\text{dilution factor})$$
$$= (13\ mg/L)(10)$$
$$= 130\ mg/L$$

**The answer is (D).**

**28.** The microorganisms are aerobic. They act upon the dissolved organic materials in the presence of oxygen. The first-stage degradation involves those microorganisms that attack carbonaceous material. Only after the completion of the first sate are the nitrogenous organic materials attacked by another variety of microorganism.

**The answer is (C).**

**29.** The ratio of peak to average daily flow rate is given in a sewage flow ratio curves figure. The *NCEES Handbook* also gives an empirical formula for the ratio, which, when $P$ is population in thousands, is

**Sewage Flow Ratio Curves**

$$\frac{\text{peak flow rate}}{\text{average flow rate}} = \frac{18 + \sqrt{P}}{4 + \sqrt{P}}$$
$$= \frac{18 + \sqrt{19}}{4 + \sqrt{19}}$$
$$= 2.7$$

Another form of this empirical formula for the ratio of peak to average daily flow rate is

$$\frac{\text{peak flow rate}}{\text{average flow rate}} = \frac{14}{4 + \sqrt{P}} + 1$$
$$= \frac{14}{4 + \sqrt{19}} + 1$$
$$= 2.7$$

**The answer is (C).**

**30.** Use the formula for an aerobic digester tank volume.

**Aerobic Digestion**

$$V = \frac{Q_i(X_i + FS_i)}{X_d(k_d P_v + 1/\theta_c)}$$

$$= \frac{\left(2.8 \ \frac{m^3}{d}\right)\left(5 \ \frac{mg}{L} + (0.5)\left(200 \ \frac{mg}{L}\right)\right)}{\left(10 \ \frac{mg}{L}\right)\left(\left(0.5 \ \frac{1}{d}\right)(0.1) + \frac{1}{2 \ d}\right)}$$

$$= 53 \ m^3$$

The answer is (D).

**31.** The required current is

$$I = (FQN/n) \times E_1/E_2$$

$$= \frac{\left(96\,485 \ \frac{C}{g\cdot\text{equivalent}}\right)\left(1 \ \frac{L}{s}\right)}{100} \times \left(0.1 \ \frac{g\cdot\text{equivalent}}{L}\right) \left(\frac{0.60}{0.98}\right)$$

$$= 59.07 \ A \quad (60 \ A)$$

The answer is (A).

**32.** From the perspective of wastewater, and in the language of chemistry, the reaction can be described as oxidation. Carbonation, although it does occur here, cannot be used for describing this type of reaction.

The answer is (D).

**33.** The problem is solved using material balances. The total material balance gives the feed to the dryer, $F$, as

$$F = P + 100$$

In this equation, $P$ is the product from the dryer. A balance over the solids gives

$$(1 - 0.63)F = (1 - 0.25)P$$

Solving for $P$ gives

$$P = \left(\frac{1 - 0.63}{1 - 0.25}\right)F = 0.50F$$

Substituting this value in the equation for the total material balance gives

$$F = 0.5F + 100 \ kg$$

Solving for $F$ gives

$$F = \frac{100 \ kg}{0.5} = 200 \ kg$$

The answer is (C).

**34.** Use a continuously stirred tank reactor (CSTR) model. [**Continuous-Stirred Tank Reactor (CSTR)**]

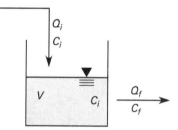

The concentration is assumed to be uniform throughout the vessel.

$$Q_i = Q_f$$

$$= \left(38\,000 \ \frac{L}{h}\right)\left(\frac{1 \ h}{60 \ min}\right)$$

$$= 630 \ L/min$$

A mass balance gives

$$Q_i C_i = Q_f C_f + kVC_f$$

Solving for $C_i$ gives

$$C_i = \frac{Q_f C_f + kVC_f}{Q_i}$$

$$= \frac{\left(630 \ \frac{L}{min}\right)C_f + \left(0.055 \ \frac{1}{min}\right)(3800 \ L)C_f}{630 \ \frac{L}{min}}$$

$$= 1.33 C_f$$

$$\text{efficiency of the system} = \frac{\text{amount removed}}{\text{amount initally present}}$$

$$= \frac{Q_i(C_i - C_f)}{Q_i C_i}$$

$$= \frac{C_i - C_f}{C_i}$$

$$= \frac{1.33 C_f - C_f}{1.33 C_f}$$

$$= 0.25$$

The answer is (B).

**35.** Use a hydraulic elements graph to obtain the solution. The ratio of depth, $d$, to diameter, $D$, must be found, as follows. [**Hydraulic-Elements Graph for Circular Sewers**]

$$\frac{d}{D} = \frac{15 \text{ cm}}{20 \text{ cm}} = 0.75$$

The hydraulic radius curve in the chart at that depth-to-diameter ratio reads 1.22, so that

$$\frac{R}{R_f} = 1.22$$

The full-flow hydraulic radius is

$$R_f = \frac{D}{4}$$

Solving for $R$ and substituting for $R_f$ yields

$$\begin{aligned} R &= 1.22 R_f \\ &= 1.22 \left(\frac{D}{4}\right) \\ &= (1.22) \left(\frac{20 \text{ cm}}{4}\right) \\ &= 6 \text{ cm} \end{aligned}$$

**The answer is (A).**

**36.** Use the Kline-McClintock equation.

$$w_R = \sqrt{\left(w_1 \frac{\partial f}{\partial x_1}\right)^2 + \left(w_2 \frac{\partial f}{\partial x_2}\right)^2 + \cdots + \left(w_n \frac{\partial f}{\partial x_n}\right)^2}$$

For the incinerability index,

$$f = IC + \frac{100 \frac{\text{kcal}}{\text{g}}}{H}$$

$$x_1 = C$$
$$x_2 = H$$
$$w_1 = w_C = (0.1)(0.02) = 0.002$$
$$w_2 = w_H = 0.06 \text{ kcal/g}$$
$$w_R = w_I$$
$$\frac{\partial f}{\partial C} = \frac{\partial C}{\partial C} = 1$$

$$\frac{\partial f}{\partial H} = \frac{\partial \left(\frac{100 \frac{\text{kcal}}{\text{g}}}{H}\right)}{\partial H} = -\frac{100 \frac{\text{kcal}}{\text{g}}}{H^2}$$

$$w_I = \sqrt{\big((0.002)(1)\big)^2 + \left(\left(0.06 \frac{\text{kcal}}{\text{g}}\right)\left(-\frac{100 \frac{\text{kcal}}{\text{g}}}{\left(2.71 \frac{\text{kcal}}{\text{g}}\right)^2}\right)\right)^2}$$

$$= 0.817 \quad (0.80)$$

This means that $I$ is $36.9 \pm 0.8$.

**The answer is (D).**

**37.** The half-life is equal to

$$t_{1/2} = 1 \text{ h} = 1/24 \text{ d}$$

The biological reaction is a Michelis-Menten type reaction, which can be approximated to be a first-order reaction. For a first-order reaction, the kinetic constant is given by

**Half-Life**

$$\begin{aligned} k &= \frac{0.693}{t_{1/2}} \\ &= \frac{0.693}{\frac{1}{24} \text{ d}} \\ &= 17/\text{d} \end{aligned}$$

**The answer is (A).**

**38.** The coefficient of variation is given by

**Data Quality Objectives (DQO) for Sampling Soils and Solids**

$$\begin{aligned} \text{CV} &= (100 * s)/\bar{x} \\ &= \frac{(100)(45 \text{ ppm})}{300 \text{ ppm}} \\ &= 15\% \end{aligned}$$

The probability of making a Type II error, $\beta$, is 10%, so power is

$$\begin{aligned} \text{power} &= 1 - \beta \\ &= 1 - 0.1 \\ &= 90\% \end{aligned}$$

From the table, the minimum detectable relative difference is 20%.

*The answer is (C).*

**39.** Use a chart of reactivity consequences. Cyanides are number 11 and organic acids are number 3. This combination results in the generation of toxic and flammable gases. [**Hazardous Waste Compatibility Chart**]

*The answer is (B).*

**40.** Use the equation for break-through time for leachate to penetrate a clay liner.

**Break-Through Time for Leachate to Penetrate a Clay Liner**

$$t = \frac{d^2 \eta}{K(d+h)}$$

$$= \frac{(0.7 \text{ m})^2 (0.2)}{\left(3 \times 10^{-7} \frac{\text{cm}}{\text{s}}\right)(0.7 \text{ m} + 2 \text{ m})\left(\frac{1 \text{ m}}{100 \text{ cm}}\right)\left(86\,400 \frac{\text{s}}{\text{d}}\right)}$$

$$= 140 \text{ d}$$

*The answer is (C).*

**41.** Use the bioconcentration formula.

**Bioconcentration Factor BCF**

$$BCF = C_{\text{org}}/C$$

Solving for $C_{\text{org}}$ gives

$$C_{\text{org}} = (BCF)C$$

$$= (10\,000)(2 \text{ ppb})\left(\frac{1 \text{ ppm}}{1000 \text{ ppb}}\right)\left(1 \frac{\frac{\text{mg}}{\text{L}}}{\text{ppm}}\right)$$

$$= 20 \text{ mg/L}$$

*The answer is (A).*

**42.** Even though treated wastewater is used for irrigation purposes, secondary and advanced treatment is required. The cost of treating the wastewater and the cost of conveying must be competitive with the corresponding costs for water treatment.

*The answer is (C).*

**43.** The terminal drift velocity of the particles does not change and neither does the gas flow rate. Use the Deutsch-Anderson equation.

**Electrostatic Precipitator Efficiency**

$$\eta = 1 - e^{(-WA/Q)}$$

Solve for $Q/W$.

$$\ln(1 - \eta) = -\frac{WA}{Q}$$

$$\frac{Q}{W} = -\frac{A}{\ln(1 - \eta)}$$

$$= \frac{600 \text{ m}^2}{\ln(1 - 0.92)}$$

$$= 240 \text{ m}^2$$

Solve the Deutsch-Anderson equation for $A$ and substitute the new required efficiency.

$$A = -\frac{Q}{W}\ln(1 - \eta)$$

$$= -(240 \text{ m}^2)\ln(1 - 0.96)$$

$$= 770 \text{ m}^2$$

*The answer is (B).*

**44.** For quartz in pulse-jet/felt baghouses, the ratio of air to cloth is 2.7. [**Air-to-Cloth Ratio for Baghouses**]

$$\frac{\text{air flow}}{\text{cloth area}} = 2.7 \frac{\frac{\text{m}^3}{\text{min}}}{\text{m}^2}$$

Solve for cloth area.

$$\text{cloth area} = \frac{\text{air flow}}{2.7 \frac{\frac{\text{m}^3}{\text{min}}}{\text{m}^2}} = \frac{2000 \frac{\text{m}^3}{\text{min}}}{2.7 \frac{\frac{\text{m}^3}{\text{min}}}{\text{m}^2}}$$

$$= 740 \text{ m}^2$$

*The answer is (A).*

**45.** The BOD just below the discharge point is the volume-weighted average of the stream BOD and the wastewater BOD.

$$C_0 = \frac{S_s Q_s + S_w Q_w}{Q_s + Q_w}$$

$$= \frac{\left(3 \frac{\text{mg}}{\text{L}}\right)\left(3 \frac{\text{m}^3}{\text{s}}\right) + \left(30 \frac{\text{mg}}{\text{L}}\right)\left(1 \frac{\text{m}^3}{\text{s}}\right)}{3 \frac{\text{m}^3}{\text{s}} + 1 \frac{\text{m}^3}{\text{s}}}$$

$$= 10 \text{ mg/L}$$

*The answer is (A).*

**46.** Smaller particles are harder to remove in a cyclone, and a higher inlet velocity reduces the cyclone's efficiency because after the drop in gas velocity, more particles will still be entrained than when the inlet velocity was lower. Particles that are less dense have more of a tendency to remain entrained and not be removed. However, increasing the length of the cone increases the cyclone's effective number of turns and efficiency.

**The answer is (D).**

**47.** Manholes are needed whenever there is a change in pipe slope.

**The answer is (B).**

**48.** Effective stack height, $H$, is given by

**Atmospheric Dispersion Modeling (Gaussian)**
$$H = h + \Delta h$$
$$= 50 \text{ m} + 10 \text{ m}$$
$$= 60 \text{ m}$$

On a cloudy night with a 4 m/s wind speed, atmospheric stability is neutral (atmospheric stability class D). Find the point on the chart for $x_{max}$ as a function of $(Cu/Q)_{max}$ for a stack whose effective height is 60 m along the line for neutral atmospheric stability. [**Atmospheric Stability Under Various Conditions**]

The corresponding distance downwind of the stack is

$$x_{max} = 1.3 \text{ km}$$

The chart for $\sigma_z$ vs. $x$ could also be used. In this case, $x_{max}$ occurs where

$$\sigma_z = \frac{H}{\sqrt{2}}$$

For this problem, $\sigma_z$ for $x_{max}$ is

$$\sigma_z = \frac{60 \text{ m}}{\sqrt{2}} = 42 \text{ m}$$

From the chart where this value intersects the neutral stability curve, this gives

$$x_{max} = 1300 \text{ m} \quad (1.3 \text{ km})$$

**The answer is (B).**

**49.** Use the NRCS rainfall-runoff formula.

**NRCS (SCS) Rainfall-Runoff**
$$Q = \frac{(P - 0.2S)^2}{P + 0.8S}$$
$$= \frac{(12 \text{ in} - (0.2)(1 \text{ in}))^2}{12 \text{ in} + (0.8)(1 \text{ in})}$$
$$= 11 \text{ in}$$

**The answer is (D).**

**50.** Use the Gaussian atmospheric dispersion modeling equation.

**Atmospheric Dispersion Modeling (Gaussian)**
$$C = \frac{Q}{2\pi u \sigma_y \sigma_z} \exp\left(-\frac{1}{2}\frac{y^2}{\sigma_y^2}\right)\left[\exp\left(-\frac{1}{2}\frac{(z-H)^2}{\sigma_x^2}\right)\right.$$
$$\left.+ \exp\left(-\frac{1}{2}\frac{(z+H)^2}{\sigma_x^2}\right)\right]$$

At ground level downwind from the stack, $y$ and $z$ are 0 m. This gives

$$C = \left(\frac{Q}{2\pi u \sigma_y \sigma_z}\right)\left[\exp\left(-\frac{1}{2}\right)\left(\frac{H^2}{\sigma_z^2}\right) + \exp\left(-\frac{1}{2}\right)\left(\frac{H^2}{\sigma_z^2}\right)\right]$$
$$= \left(\frac{Q}{\pi u \sigma_y \sigma_z}\right)\exp\left(-\frac{1}{2}\right)\left(\frac{H^2}{\sigma_z^2}\right)$$

The atmospheric stability on a clear summer day with the sun higher than 60° above the horizon and a wind speed of 4 m/s is moderately unstable (atmospheric stability class B). The values (from the graphs in the *NCEES Handbook*) for $\sigma_y$ and $\sigma_z$ at 200 m downwind of the stack are 33 m and 22 m, respectively. [**Atmospheric Stability Under Various Conditions**]

Thus,

$$C = \left(\frac{60 \frac{\text{g}}{\text{s}}}{\pi\left(7 \frac{\text{m}}{\text{s}}\right)(33 \text{ m})(22 \text{ m})}\right)\exp\left(-\frac{1}{2}\right)\left(\frac{(20 \text{ m})^2}{(22 \text{ m})^2}\right)$$
$$= 2.5 \times 10^{-3} \text{ g/m}^3$$

Relative to the given nighttime concentration, this value is

$$\frac{2.5 \times 10^{-3} \frac{\text{g}}{\text{m}^3}}{1.5 \times 10^{-4} \frac{\text{g}}{\text{m}^3}} = 17 \text{ times higher}$$

**The answer is (C).**

**51.** The quantity of fuel oil required is the heat deficiency divided by the fuel's heating value.

$$\text{quantity of fuel required} = \frac{\text{heat deficiency}}{\text{fuel heating value}}$$

$$= \frac{\left(4020 \ \frac{\text{MJ}}{\text{h}}\right)\left(1000 \ \frac{\text{kJ}}{\text{MJ}}\right)}{15\,200 \ \frac{\text{kJ}}{\text{L}}}$$

$$= 260 \ \text{L/h}$$

**The answer is (D).**

**52.** The ion product of water is

$$K_w = [\text{H}_3\text{O}^+][\text{OH}^-] = 10^{-14} \ \text{M}^2$$

NaOH is a strong base and completely dissociates. The contribution to [OH$^-$] from water is small compared to the concentration of [OH$^-$] from NaOH, so

$$[\text{OH}^-] = \frac{1 \ \text{mol}}{10 \ \text{L}} = 0.1 \ \text{M}$$

Solve for [H$_3$O$^+$] to get

$$[\text{H}_3\text{O}^+] = \frac{10^{-14} \ \text{M}^2}{0.1 \ \text{M}} = 10^{-13} \ \text{M}$$

$$\text{pH} = -\log \ [\text{H}_3\text{O}^+] = -\log 10^{-13} = 13$$

**The answer is (C).**

**53.** Underground storage tank regulations are primarily intended to protect groundwater from contamination due to leaking tanks.

**The answer is (B).**

**54.** Let $t_D$ be the time it takes for the dissolved oxygen concentration to reach 7.3 mg/L. If DO represents dissolved oxygen concentration, $D$ is given by

$$D = \text{saturation DO} - \text{actual DO}$$

In winter, at $t = 0$ h,

$$D_{0\text{h, winter}} = 11.3 \ \frac{\text{mg}}{\text{L}} - 5.8 \ \frac{\text{mg}}{\text{L}} = 5.5 \ \text{mg/L}$$

At $t = t_d$,

$$D_{t_d, \text{winter}} = 11.3 \ \frac{\text{mg}}{\text{L}} - 7.3 \ \frac{\text{mg}}{\text{L}} = 4.0 \ \text{mg/L}$$

The reaeration coefficient in winter is

$$k(T) = \left(\frac{0.44}{\text{h}}\right)(1.02^{(T-20\circ\text{C})})$$

$$= \left(\frac{0.44}{\text{h}}\right)(1.02^{(10\circ\text{C}-20\circ\text{C})})$$

$$= 0.36/\text{h}$$

The rate of transfer of oxygen into water is given by

**First-Order Irreversible Reaction**

$$-dC_A/dt = kC_A$$

$$\frac{dD}{dt} = -kD$$

Integrating and solving for $t_D$ gives

$$\int_{5.5 \ \frac{\text{mg}}{\text{L}}}^{4.0 \ \frac{\text{mg}}{\text{L}}} \frac{dD}{D} = -\frac{0.36}{\text{h}} \int_{0 \ \text{h}}^{t_D} dt$$

$$\ln D \Big|_{5.5 \ \frac{\text{mg}}{\text{L}}}^{4.0 \ \frac{\text{mg}}{\text{L}}} = -\left(\frac{0.36}{\text{h}}\right) t \Big|_{0 \ \text{h}}^{t_D}$$

$$\ln 4.0 \ \frac{\text{mg}}{\text{L}} - \ln 5.5 \ \frac{\text{mg}}{\text{L}} = -\left(\frac{0.36}{\text{h}}\right) t_D + 0 \ \text{h}$$

$$t_D = -\frac{\ln \dfrac{4.0 \ \frac{\text{mg}}{\text{L}}}{5.5 \ \frac{\text{mg}}{\text{L}}}}{\dfrac{0.36}{\text{h}}}$$

$$= 0.88 \ \text{h}$$

This is 3.8 times less than the time it takes to recover to a dissolved oxygen concentration of 7.3 mg/L in summer.

**The answer is (A).**

**55.** Exposure is inversely proportional to the square of the distance from the source.

**Half-Life: Inverse Square Law**

$$\frac{I_1}{I_2} = \frac{(R_2)^2}{(R_1)^2}$$

$$\frac{I_{8\text{m}}}{I_{4\text{m}}} = \frac{\dfrac{1}{64 \ \text{m}^2}}{\dfrac{1}{16 \ \text{m}^2}} = 1/4$$

Thus, doubling the distance from the source reduces exposure by a factor of 4.

**The answer is (B).**

**56.** The volume of water in the lake is the lake's area times its average depth, or

$$V = Ad$$
$$= (3.0 \text{ km}^2)(5 \text{ m})\left(1000 \frac{\text{m}}{\text{km}}\right)^2$$
$$= 15 \times 10^6 \text{ m}^3$$

Rainfall flow into the lake is

$$Q_{\text{rainfall}} = A\left(0.7 \frac{\text{m}}{\text{yr}}\right)$$
$$= (3.0 \text{ km}^2)\left(0.7 \frac{\text{m}}{\text{yr}}\right)\left(\frac{1 \text{ yr}}{365 \text{ d}}\right)\left(\frac{1 \text{ d}}{86\,400 \text{ s}}\right)$$
$$\times \left(1000 \frac{\text{m}}{\text{km}}\right)^2$$
$$= 0.067 \text{ m}^3/\text{s}$$

At steady state, a phosphorus mass balance around the lake gives

$$Q_{\text{rainfall}}C_{\text{rainfall}} + Q_{\text{inlet stream}}C_{\text{inlet stream}}$$
$$- Q_{\text{outlet stream}}C_{\text{lake}} - k_{\text{precipitation}}C_{\text{lake}}V = 0 \text{ g/s}$$

Solve for $C_{\text{lake}}$.

$$C_{\text{lake}} = \frac{Q_{\text{rainfall}}C_{\text{rainfall}} + Q_{\text{inlet stream}}C_{\text{inlet stream}}}{Q_{\text{outlet stream}} + k_{\text{precipitation}}V}$$

$$= \frac{\left(0.067 \frac{\text{m}^3}{\text{s}}\right)\left(0.01 \frac{\text{mg}}{\text{L}}\right) + \left(0.2 \frac{\text{m}^3}{\text{s}}\right)\left(0.1 \frac{\text{mg}}{\text{L}}\right)}{0.27 \frac{\text{m}^3}{\text{s}} + \left(0.005 \frac{1}{\text{d}}\right)(15 \times 10^6 \text{ m}^3)\left(\frac{1 \text{ d}}{86\,400 \text{ s}}\right)}$$

$$= 0.018 \text{ mg/L}$$

**The answer is (A).**

**57.** The molecular weight of ammonia is

$$(1)\left(14 \frac{\text{g}}{\text{mol}}\right) + (4)\left(1 \frac{\text{g}}{\text{mol}}\right) = 18 \text{ g/mol}$$

The molecular weight of oxygen is

$$(2)\left(16 \frac{\text{g}}{\text{mol}}\right) = 32 \text{ g/mol}$$

From the chemical equation, for every mole of ammonia, 1.731 moles of oxygen are consumed. Thus, the amount of oxygen consumed when 30 kg of ammonia are released is

$$(30 \text{ kg})\left(\frac{1 \text{ mol NH}_4}{18 \text{ g NH}_4}\right)$$
$$\times \left(1.731 \frac{\text{mol O}_2}{\text{mol NH}_4}\right)\left(32 \frac{\text{g O}_2}{\text{mol O}_2}\right) = 92 \text{ kg O}_2$$

**The answer is (D).**

**58.** Mixed waste is defined as a waste mixture that contains both radioactive materials subject to the AEA and a hazardous waste component regulated under the RCRA.

**The answer is (A).**

**59.** The corrosivity of water in drinking water systems is important from a human health standpoint because if the water is too corrosive there is a potential for unhealthy levels of lead and copper in the distribution network to contaminate the water by the time it reaches the tap. An important parameter in selecting appropriate corrosion control strategies is the water's dissolved inorganic carbon (DIC).

First, write the expressions for the dissociation constants. The first dissociation reaction of carbonate is

$$H_2CO_3 \rightleftharpoons HCO_3^- + H^+$$

The equilibrium constant for the first dissociation reaction is

*Acids, Bases, and pH*

$$K_1 = \frac{[HCO_3^-][H^+]}{[H_2CO_3]}$$

The second dissociation reaction is

$$HCO_3^- \rightleftharpoons CO_3^{2-} + H^+$$

The equilibrium constant for the second dissociation reaction is

*Acids, Bases, and pH*

$$K_2 = \frac{[CO_3^{2-}][H^+]}{[HCO_3^-]}$$

The self-ionization constant of water is

$$K_w = [OH^-][H^+]$$

pH is related to the concentration of hydrogen ions in mol/L.

### Acids, Bases, and pH

$$\text{pH} = \log_{10}\left(\frac{1}{[\text{H}^+]}\right)$$
$$= -\log[\text{H}^+]$$

Solve for the concentration of hydrogen ions.

$$10^{-\text{pH}} = [\text{H}^+]$$

Find $[\text{H}_2\text{CO}_3]$ and $[\text{CO}_3^{2-}]$ in terms of $[\text{HCO}_3^-]$.

$$[\text{H}_2\text{CO}_3] = \frac{[\text{HCO}_3^-][\text{H}^+]}{K_1}$$

$$[\text{CO}_3^{2-}] = \frac{K_2[\text{HCO}_3^-]}{[\text{H}^+]}$$

Solve for [OH].

$$[\text{OH}^-] = \frac{K_w}{[\text{H}^+]}$$

Substitute into the equation for alkalinity and solve for $[\text{HCO}_3^-]$.

$$\text{Alk} = [\text{HCO}_3^-] + 2[\text{CO}_3^{2-}] + [\text{OH}^-] - [\text{H}^+]$$
$$= [\text{HCO}_3^-] + \frac{2K_2[\text{HCO}_3^-]}{[\text{H}^+]} + \frac{K_w}{[\text{H}^+]} - [\text{H}^+]$$

$$\text{Alk} - \frac{K_w}{[\text{H}^+]} + [\text{H}^+] = [\text{HCO}_3^-]\left(1 + \frac{2K_2}{[\text{H}^+]}\right)$$

$$[\text{HCO}_3^-] = \frac{\text{Alk} - \dfrac{K_w}{[\text{H}^+]} + [\text{H}^+]}{1 + \dfrac{2K_2}{[\text{H}^+]}}$$

The alkalinity must be converted from units of mg/L of $\text{CaCO}_3$ to equivalents, and expressed in units of mol/L so that it is congruent with the units in the preceding equation.

$$\left(\frac{160 \text{ mg CaCO}_3}{\text{L}}\right)\left(\frac{1 \text{ mol CaCO}_3}{100 \text{ g CaCO}_3}\right)$$
$$\times \left(\frac{1 \text{ g}}{1000 \text{ mg}}\right)\left(2 \ \frac{\text{molar equivalents}}{\text{mol CO}_3^{2-}}\right) = 0.0032 \text{ mol/L}$$

Substitute, solve for DIC, and convert from mol carbonates/L to mg C/L.

$$\text{DIC} = [\text{H}_2\text{CO}_3] + [\text{HCO}_3^-] + [\text{CO}_3^{2-}]$$
$$= \frac{[\text{HCO}_3^-][\text{H}^+]}{K_1} + [\text{HCO}_3^-] + \frac{K_2[\text{HCO}_3^-]}{[\text{H}^+]}$$
$$= [\text{HCO}_3^-]\left(\frac{[\text{H}^+]}{K_1} + 1 + \frac{K_2}{[\text{H}^+]}\right)$$
$$= \left(\frac{\text{Alk} - \dfrac{K_w}{[\text{H}^+]} + [\text{H}^+]}{1 + \dfrac{2K_2}{[\text{H}^+]}}\right)\left(\frac{[\text{H}^+]}{K_1} + 1 + \frac{K_2}{[\text{H}^+]}\right)$$

$$= \left(\frac{0.0032 \ \dfrac{\text{mol}}{\text{L}} - \dfrac{10^{-14} \ \dfrac{\text{mol}^2}{\text{L}^2}}{10^{-6.6} \ \dfrac{\text{mol}}{\text{L}}} + 10^{-6.6} \ \dfrac{\text{mol}}{\text{L}}}{1 + \dfrac{(2)\left(10^{-10.3} \ \dfrac{\text{mol}}{\text{L}}\right)}{10^{-6.6} \ \dfrac{\text{mol}}{\text{L}}}}\right)$$

$$\times \left(\frac{10^{-6.6} \ \dfrac{\text{mol}}{\text{L}}}{10^{-6.4} \ \dfrac{\text{mol}}{\text{L}}} + 1 + \frac{10^{-10.3} \ \dfrac{\text{mol}}{\text{L}}}{10^{-6.6} \ \dfrac{\text{mol}}{\text{L}}}\right)$$

$$\times \left(\frac{12 \text{ g C}}{1 \text{ mol carbonates}}\right)\left(1000 \ \frac{\text{mg}}{\text{g}}\right)$$

$$= 63 \text{ mg C/L}$$

**The answer is (B).**

**60.** The steady-state atmospheric residence time is the total amount of material in the atmosphere divided by the steady-state flux of material either into or out of the atmosphere, or

$$\tau = \frac{M}{F}$$

$$= \frac{(4.0 \text{ Tg})\left(1 \times 10^6 \ \dfrac{\text{Mg}}{\text{Tg}}\right)}{\left(2.7 \ \dfrac{\text{Mg}}{\text{s}} + 3.2 \ \dfrac{\text{Mg}}{\text{s}}\right)\left(86\,400 \ \dfrac{\text{s}}{\text{d}}\right)}$$

$$= 7.8 \text{ d}$$

**The answer is (B).**

**61.** The equation needs to be put in the form $f(x) = 0$.

$$f(x) = x^3 + 2x - 5 = 0$$

This can also be written as

$$\frac{df}{dx} = f^1(x) = 3x^2 + 2$$

The equation for Newton's method of root extraction is

$$X_{n+1} = X_n - \frac{f(X_n)}{f^{11}(X_n)}$$

**Newton's Method for Root Extraction**

$$a^{j+1} = a^j - \frac{f(x)}{\frac{df(x)}{dx}}\bigg|_{x=a^j}$$

The equation written in this form is

$$a^{j+1} = \frac{a_j^3 + 2a_j - 5}{3a_j^2 + 2}$$

Solve the equation for the possible values of $n$ and tabulate the results. Choose an initial value of $n = 0$ and calculate; then calculate the answer for $n = 1$ using the found value; then calculate the value for $n = 2$ using that found value. Continue this process until the difference between successive iterations is less than 5%.

| $j$ | $a^j$ | $a^{j+1}$ |
|---|---|---|
| 0 | 1.1 | 1.36 |
| 1 | 1.36 | 1.329 |
| 2 | 1.329 | 1.328 |

$a$ is approximately equal to 1.32.

**The answer is (D).**

**62.** The optimal point of the function $f(x)$ is the point $x$ at which $df/dx = 0$. The optimal point becomes a maximum if the equation shown is true at that point.

$$\frac{d^2f}{dx^2} < 0$$

The optimal point becomes a minimum if the equation shown is true at that point.

$$\frac{d^2f}{dx^2} > 0$$

The equation for $df/dx$ in this case is

$$\frac{df}{dx} = f^1(x) = 2x + 4$$

An optimal point occurs when $f^1(x) = 0 = 2x + 4$; in other words, when $x = -4/2 = -2$. The value of $d^2f/dx^2$ for this case is

$$\frac{d^2f}{dx^2} = \left(\frac{d}{dx}\right)\left(\frac{df}{dx}\right) = 2$$

This value is positive. Therefore, there is a minimum at $x = -2$.

**The answer is (A).**

**63.** The equation for the velocity of the particle is

**Particle Rectilinear Motion**

$$v = \frac{ds}{dt}$$
$$= 6t^{2-1}$$

Therefore, the equation for the acceleration is

$$a = \frac{d^2s}{dt^2} = 12t$$

Solve for $t$ when the acceleration is 24 m/s².

$$a = \frac{d^2s}{dt^2} = 12t$$
$$t = \frac{a}{12}$$
$$= \frac{24\ \frac{m}{s^2}}{12}$$
$$= 2\ \text{sec}$$

The displacement at $t = 2$ sec is

$$s = 2t^3 - t$$
$$= (2)(2\ \text{sec})^3 - 2\ \text{sec}$$
$$= 14\ \text{m}$$

**The answer is (D).**

**64.** The equation for an ellipse centered at (0, 0) is

**Conic Sections**

$$\frac{(x-h)^2}{a^2} + \frac{(y-k)^2}{b^2} = 1$$

$$\frac{x^2}{4} + \frac{y^2}{9} = 1$$

The equation shown in the problem statement is thus an ellipse with the center at (2, 3). The major axis is $x=2$, which is perpendicular to the $x$-axis at $x=2$. The minor axis is $y=3$, which is perpendicular to the $y$-axis at $y=3$.

**The answer is (A).**

**65.** The second-order differential is defined as

$$\frac{d^2y}{dx^2} = \left(\frac{d}{dx}\right)\left(\frac{dy}{dx}\right)$$

The equation given in the problem statement can be rewritten as

$$\left(\frac{d}{dx}\right)\left(\frac{dy}{dx}\right) = 2xe^x + 3\cos x$$

Integrating the outer differential on the left side and the expression the right side, the equation becomes

$$\int \left(\frac{d}{dx}\right)\left(\frac{dy}{dx}\right)dx = \int (2xe^x + 3\cos x)\,dx$$

Integrate the right side by by parts, with $c_1$ as the constant of integration.

$$\frac{dy}{dx} = 2(xe^x - e^x) + 3\sin x + c_1$$

Integrating again, where $c_2$ is the second constant of integration,

$$\int \left(\frac{dy}{dx}\right)dx = \int \left((2)(xe^x - e^x) + 3\sin x + c_1\right)dx$$

$$y = 2xe^x - 2e^x - 2e^x - 3\cos x + c_1 x + c_2$$
$$= 2xe^x - 4e^x - 3\cos x + c_1 x + c_2$$

**The answer is (B).**

**66.** The heat loss is

**Heat Exchangers**

$$\dot{Q} = UA\Delta T_{lm}$$
$$= \left(7.50\,\frac{\text{W}}{\text{m}^2\cdot\text{K}}\right)(1.2\text{ m}^2)(85°\text{C} - 20°\text{C})$$
$$= 585\text{ J/s}$$

The heat loss can also be expressed as

$$Q = \dot{m}C_p \Delta T_{\text{loss}}$$

Rearrange the equation for heat loss as a function of temperature loss to find the temperature loss in the fluid.

$$Q = \dot{m}C_p \Delta T_{\text{loss}} = 585\text{ J/s}$$

$$\Delta T_{\text{loss}} = \frac{Q}{\dot{m}C_p}$$

$$= \frac{585\,\frac{\text{J}}{\text{s}}}{\left(15\,\frac{\text{kg}}{\text{min}}\right)\left(\frac{1\text{ min}}{60\text{ sec}}\right)\left(10^3\,\frac{\text{g}}{\text{kg}}\right)\left(0.95\,\frac{\text{J}}{\text{g}\cdot°\text{C}}\right)}$$

$$= 2.46°\text{C}\quad (2.4°\text{C})$$

**The answer is (A).**

**67.** The mean is the average of all the values in the set. **[Dispersion, Mean, Median, And Mode Values]**

The mean dissolved oxygen (DO) is

$$\frac{\begin{array}{l}8.0\,\frac{\text{mg}}{\text{L}} + 7.8\,\frac{\text{mg}}{\text{L}} + 7.5\,\frac{\text{mg}}{\text{L}} + 7.0\,\frac{\text{mg}}{\text{L}} + 6.0\,\frac{\text{mg}}{\text{L}} \\ +5.4\,\frac{\text{mg}}{\text{L}} + 5.0\,\frac{\text{mg}}{\text{L}} + 4.8\,\frac{\text{mg}}{\text{L}} + 4.7\,\frac{\text{mg}}{\text{L}}\end{array}}{9}$$

$$= 6.244\text{ mg/L}\quad (6.2\text{ mg/L})$$

**The answer is (B).**

**68.** The median value is the value that is numerically in the middle of the provided values. There are an equal number of data points above and below 6.0 mg/L, so 6.0 mg/L is the median value.

**The answer is (C).**

**69.** At a pressure of 1 atm, the gas can be assumed to be an ideal gas. Since the tank system is completely insulated, the external heat transfer is equal to 0.

The tank system is made of rigid materials, so the external work done is equal to 0. By the first law of thermodynamics, **[First Law of Thermodynamics]**

$$\Delta U = R - W = 0$$

Since $U$ is constant, from the definition of an ideal gas, the temperature of the gas is constant.

**The answer is (D).**

**70.** The equation for fracture toughness is

**Properties of Materials: Mechanical**
$$K_{IC} = Y\sigma\sqrt{\pi a}$$

The fracture toughness of 4340 steel is 46 MPa·m$^{1/2}$. **[Representative Values of Fracture Toughness]**

For interior cracks, the width of the crack is $2a$.

$$2 \text{ mm} = 2a$$

Solve for $a$.

$$a = \frac{2 \text{ mm}}{2} = 1 \text{ mm}$$

Rearrange the equation for fracture toughness to solve for the engineering stress and substitute values.

$$\sigma = \frac{K_{IC}}{Y\sqrt{\pi a}}$$

$$= \frac{46 \text{ MPa} \cdot \text{m}^{\frac{1}{2}}}{1\sqrt{\pi (1 \text{ mm})\left(\dfrac{1 \text{ m}}{1000 \text{ mm}}\right)}}$$

$$= 820 \text{ MPa}$$

**The answer is (D).**

**71.** Use 1 the gas mixture as a basis for calculation.

The mass of nitrogen ($N_2$) in 100 g of the gas mixture is 40 g. From the periodic table of elements, the atomic mass of nitrogen is approximately 14 g/mol. [**Periodic Table of Elements**].

The molecular weight of $N_2$ is

$$\text{MW}_{N_2} = \left(14 \text{ }\frac{\text{g}}{\text{mol}}\right)(2) = 28 \text{ g/mol}$$

The quantity of moles of nitrogen in the mixture is

$$n_{N_2} = \frac{m}{\text{MW}_{N_2}} = \frac{40 \text{ g}}{28 \text{ }\frac{\text{g}}{\text{mol}}} = 1.42 \text{ mol}$$

The mass of oxygen ($O_2$) in 100 g of the gas mixture is 40 g. From the periodic table of elements, the atomic mass of oxygen is approximately 16 g/mol. [**Periodic Table of Elements**]

The molecular weight of $O_2$ is

$$\text{MW}_{O_2} = \left(16 \text{ }\frac{\text{g}}{\text{mol}}\right)(2) = 32 \text{ g/mol}$$

The quantity of moles of oxygen in the mixture is

$$n_{O_2} = \frac{m}{\text{MW}_{O_2}} = \frac{40 \text{ g}}{32 \text{ mol}} = 1.25 \text{ mol}$$

The mass of carbon dioxide ($CO_2$) in 100 g of the gas mixture is 20 g. From the periodic table of the elements, the atomic mass of carbon is approximately 12 g/mol, and the atomic mass of oxygen is approximately 16 g/mol. [**Periodic Table of Elements**]

The molecular weight of carbon monoxide is

$$\text{MW}_{CO} = 12 \text{ }\frac{\text{g}}{\text{mol}} + 16 \text{ }\frac{\text{g}}{\text{mol}} = 28 \text{ g/mol}$$

The quantity of moles of carbon monoxide in the mixture is

$$n_{CO} = \frac{m}{\text{MW}_{CO}} = \frac{20 \text{ g}}{28 \text{ }\frac{\text{g}}{\text{mol}}} = 0.714 \text{ mol}$$

The total quantity of moles in the mixture is

$$1.42 \text{ mol} + 1.25 \text{ mol} + 0.714 \text{ mol} = 3.384 \text{ mol}$$

The mole fraction of nitrogen in the mixture is

$$x_{N_2} = \frac{n_{N_2}}{n_{\text{total}}} = \frac{1.42 \text{ mol}}{3.384 \text{ mol}}$$

$$= 0.4196$$

The partial pressure of nitrogen in the mixture is

$$P_{N_2} = x_{N_2} P_{\text{total}} = (0.4196)(2 \text{ atm})$$

$$= 0.8392 \text{ atm} \quad (0.84 \text{ atm})$$

**The answer is (C).**

**72.** Use 100 moles of the fluid as the basis of the calculation.

100 moles of the fluid contains 80 mol of nitrogen ($N_2$), 1 mol of oxygen ($O_2$), and 1 mol of carbon dioxide ($CO_2$).

If $x$ is the quantity of moles of $N_2$ added, then the total quantity of moles in the mixture is 100 mol + $x$. The proportion of carbon dioxide in the mixture is then

$$\frac{1}{100 + x}$$

For the mixture after the addition to be 0.5% carbon dioxide, then the relationship shown must be true.

$$\frac{1}{100 \text{ mol} + x} = \frac{0.5}{100 \text{ mol}}$$

Rearrange and solve for $x$.

$$\frac{1}{100 \text{ mol} + x} = \frac{0.5}{100 \text{ mol}}$$

$$100 \text{ mol} = 50 \text{ mol} + 0.5x$$

$$x = \frac{100 \text{ mol} - 50 \text{ mol}}{0.5}$$

$$= 100 \text{ mol}$$

The quantity of moles of nitrogen in the mixture after the addition is

$$80 \text{ mol} + 100 \text{ mol} = 180 \text{ mol}$$

The partial pressure of nitrogen in the mixture after the addition is

$$P_{N_2} = x_{N_2} P_{\text{total}} = \left(\frac{180 \text{ mol}}{200 \text{ mol}}\right)(1.2 \text{ atm})$$

$$= 1.08 \text{ atm}$$

**The answer is (B).**

**73.** The layers of the slab can be treated as a series of resistors for the heat energy. Resistance is the inverse of conductance, and for a series of resistances, the total resistance is the sum of the resistances. [**Thermal Resistance ($R$)**]

The total resistance of the slab is

$$R_{\text{total}} = \frac{1}{k_1} + \frac{1}{k_2}$$

$$= \frac{1}{1.5 \ \frac{\text{kJ}}{\text{m}^2 \cdot \text{K}}} + \frac{1}{2.5 \ \frac{\text{kJ}}{\text{m}^2 \cdot \text{K}}}$$

$$= 1.067 \text{ m}^2 \cdot \text{K/kJ}$$

The total conductance of the slab is

$$C = \frac{1}{R} = \frac{1}{1.067 \ \frac{\text{m}^2 \cdot \text{K}}{\text{kJ}}} = 0.9375 \text{ kJ/m}^2 \cdot \text{K}$$

The heat transmitted is

$$\text{total heat} = \frac{C(T_2 - T_1)}{N_{\text{layers}}}$$

$$= \frac{\left(0.9375 \ \frac{\text{kJ}}{\text{m}^2 \cdot \text{K}}\right)(1 \text{ m}^2)(100°\text{C})}{2}$$

$$= 46.875 \text{ kJ/m}^2 \quad (48 \text{ kJ/m}^2)$$

**The answer is (A).**

**74.** The rules of professional conduct are binding upon every engineering licensee. Among other things, these rules state that "Licensees shall notify their employer or client and such other authority as may be appropriate when their professional judgment is overruled under circumstances in which the health, safety, or welfare of the public is endangered."

Model Rules, Section 240.15, Rules of Professional Conduct

**The answer is (A).**

**75.** Since the job offered is a managerial position and does not involve technical work, a license in any one discipline is sufficient. No additional licensure is needed.

**The answer is (B).**

**76.** The reaction rate constant, $k$, is related to the activation energy, $\Delta E$, as shown.

$$k = k_o \frac{-\Delta E}{RT}$$

$k_o$ is a constant, $\Delta E$ is the activation energy, and $T$ is the absolute temperature.

If $k_1$ is the constant at temperature $T_1$, and $k_2$ is the rate constant at temperature $T_2$,

$$\log_{10}\left(\frac{K_2}{K_1}\right) = \frac{-\Delta E}{2.303R}\left(\frac{1}{T_2} - \frac{1}{T_1}\right)$$

Find $T_1$ and $T_2$.

$$T_1 = 20°\text{C} + 273° = 293\text{K}$$

$$T_2 = 70°\text{C} + 273° = 343\text{K}$$

$K_2 = 10K_1$, so the activation energy of the reaction is

$$\log_{10}\left(\frac{K_2}{K_1}\right) = \log_{10}\left(\frac{10K_1}{K_1}\right) = \frac{-\Delta E}{2.303R}\left(\frac{1}{T_2} - \frac{1}{T_1}\right)$$

$$\Delta E = \frac{-\log_{10}\left(\frac{10K_1}{K_1}\right)(2.303R)}{\frac{1}{T_2} - \frac{1}{T_1}}$$

$$= \frac{-\log_{10}(10)(2.303)\left(1.987 \ \frac{\text{cal}}{\text{mol} \cdot \text{K}}\right)}{\frac{1}{343\text{K}} - \frac{1}{293\text{K}}}$$

$$= 9.207 \text{ kcal/mol} \quad (9.2 \text{ kcal/mol})$$

**The answer is (C).**

**77.** Check that the sum of the probabilities is equal to 1.00.

$$\sum p = 0.15 + 0.15 + 0.20 + 0.20 + 0.30 = 1.00$$

The expected value is equal to the sum of the products of the net annual worth for each scenario and the probability that scenario will occur.

**Expected Values**

$$E[X] = \sum x_k f(x_k) = \sum p_n A_{\text{net},n}$$
$$= (0.15)(\$10{,}000{,}000)$$
$$+ (0.15)(\$12{,}000{,}000)$$
$$+ (0.20)(\$15{,}000{,}000)$$
$$+ (0.20)(\$18{,}000{,}000)$$
$$+ (0.30)(\$25{,}000{,}000)$$
$$= \$17{,}400{,}000 \quad (\$17{,}000{,}000)$$

**The answer is (C).**

**78.** The equation for the volumetric flow rate is

**Venturi Meters**

$$Q = \frac{C_v A_2}{\sqrt{1-(A_2/A_1)^2}} \sqrt{2g\left(\frac{P_1}{\gamma} + z_1 - \frac{P_2}{\gamma} - z_2\right)}$$

This can be simplified to

$$Q = \frac{C_v A_2}{\sqrt{1-(A_2/A_1)^2}} \sqrt{\frac{2(P_1-P_2)}{\rho}}$$

The ratio of the areas is

$$\frac{A_2}{A_1} = \left(\frac{D_2}{D_1}\right)^2 = \left(\frac{1.5 \text{ in}}{3.0 \text{ in}}\right)^2 = 0.25$$

The area of the throat in units of square meters is

$$A_2 = \frac{\pi D_2^2}{4}$$
$$= \left(\frac{\pi}{4}\right)\left(\frac{1.5 \text{ in}}{12 \frac{\text{in}}{\text{ft}}}\right)^2 \left(30 \frac{\text{cm}}{\text{ft}}\right)^2 \left(\frac{1 \text{ m}}{100 \text{ cm}}\right)^2$$
$$= 0.0011 \text{ m}^2$$

The coefficient $C_v = 0.98$ for a rounded venturi meter.

Convert the pressure drop to units of $N/m^2$.

$$(3 \text{ mm Hg})\left(\frac{\text{atm}}{760 \text{ mm Hg}}\right)$$
$$\times \left(\frac{1.013 \times 10^5 \text{ N}}{\text{atm} \cdot \text{m}^2}\right) = 399 \text{ N/m}^2$$

The volumetric flow rate is

$$Q = \left(\frac{(0.98)(0.0011 \text{ m}^2)}{\sqrt{1-(0.25)^2}}\right)\sqrt{\frac{(2)\left(399 \frac{\text{N}}{\text{m}^2}\right)}{\left(997 \frac{\text{kg}}{\text{m}^3}\right)}}$$
$$= 0.0009 \text{ m}^3/\text{s}$$

**The answer is (A).**

**79.** From Bernoulli's equation for two points on the surface of a liquid, both before and after the change in cross section the pressure on the top of the liquid is equal to the pressure of the prevailing atmosphere, and therefore, the same at both points. **[Energy Equation]**

The velocity before the change in cross section will be less than the velocity after the change, due to the large cross section and due to volumetric flow remaining constant at steady flow. Since there is no pump work, and friction losses can be considered negligible, the height of the liquid will decrease in order to maintain a constant total head.

**The answer is (B).**

**80.** One environmental expert must be on the committee, so one more environmental experts must be selected from among the $6 - 1 = 5$ remaining experts. The number of ways the environmental expert can be selected is

$$\frac{5!}{1!(5-1)!} = \frac{(5)(4)(3)(2)(1)}{(1)((4)(3)(2)(1))} = 5$$

The number of ways three lay persons can be selected out of the six eligible lay persons is

$$\frac{6!}{3!(6-3)!} = \frac{(6)(5)(4)(3)(2)(1)}{(3)(2)(1)((3)(2)(1))} = 20$$

The total number of ways the subcommittee can be formed is

$$(5)(20) = 100$$

**The answer is (A).**

**81.** The equation for the velocity at the given point is

**Pitot Tube**
$$v = \sqrt{2g(P_0 - P_s)/\gamma}$$

$P_0$ is the stagnation pressure in units of $N/m^2$, $P_s$ is the static pressure in units of $N/m^2$, and $\gamma$ is the specific weight of the fluid.

The difference between the stagnation head and the static head is $1.6 \text{ m} - 1 \text{ m} = 0.6 \text{ m}$. Convert the difference between the heads to pressure.

$$\left(\frac{0.6 \text{ m}}{\left(33.90 \frac{\text{ft water}}{\text{atm}}\right)\left(0.3 \frac{\text{m}}{\text{ft}}\right)}\right) \times (1 \text{ atm})\left(1.013 \times 10^5 \frac{\text{Pa}}{\text{atm}}\right) = 5957.5 \text{ N/m}^2$$

The specific weight of water is $9.807 \times 10^3 \text{ kN/m}^3$, or $9.807 \times 10^3 \text{ (kg/m}^3)\cdot(\text{m/s}^2)$. The velocity is

$$v = \frac{(2)\left(9.81 \frac{\text{m}}{\text{s}^2}\right)\left(5957.5 \frac{\text{N}}{\text{m}^2}\right)}{9.807 \times 10^3 \frac{\text{kg}}{\text{m}^3} \cdot \frac{\text{m}}{\text{s}^2}}$$

$$= 3.46 \text{ m/s} \quad (3.5 \text{ m/s})$$

**The answer is (B).**

**82.** The expected cost of the project after a new law is passed is equal to the sum of the products of the probability of each law passing and the cost of the project under that law.

$$\text{expected cost} = \sum p(\text{cost})$$
$$= (0.3)(\$10,000,0000)$$
$$\quad + (0.5)(\$12,000,0000)$$
$$\quad + (0.2)(\$24,000,0000)$$
$$= \$13,800,000$$

The current year is year 1, so treat the year after the law is passed as year 2. From interest rate tables, for an interest rate of 6% and a time of 2 yr, the cost factor $P/F$ is 0.8900. [**Interest Rate Tables**]

The present worth of a cost of \$13,800,000 in year 2 is

$$(\$13,800,000)\left(\frac{P}{F}\right)\bigg|_{t=2 \text{ yr}}^{6\%} = (\$13,800,000)(0.89)$$
$$= \$12,282,000$$

The expected change in the cost if the project is started after the new law passes is

$$\frac{\$12,282,000 - \$10,000,000}{\$10,000,000} = 22.82\% \quad (23\%)$$

**The answer is (C).**

**83.** Find the equivalent present worth of the costs. From interest rate tables, the $P/F$ cost factor for an interest rate of 4% and a time of 10 yr is 0.6756. [**Interest Rate Tables**]

$$P = \text{initial cost} + F\left(\frac{P}{F}\right)\bigg|_{10\text{yr}}^{4\%}$$
$$= \$10,000,000 + (\$5,000,000)(0.6756)$$
$$= \$13,378,000$$

Find the present worth of the benefits. From interest rate tables, the $P/A$ cost factor for an interest rate of 4% and a time of 10 yr is 8.1109. [**Interest Rate Tables**]

$$P = B\left(\frac{P}{A}\right)\bigg|_{10\text{yr}}^{4\%}$$
$$= (\$30,000,000)(8.1109)$$
$$= \$243,327,000$$

The benefit-cost ratio is

$$B/C = \frac{\$243,327,000}{\$13,378,000} = 18.188 \quad (18.0)$$

**The answer is (C).**

**84.** The value of the $A/P$ cost factor for an annual benefit of \$1,000,000 and an initial investment of \$5,000,000 is

$$\frac{A}{P} = \frac{\$1,000,000}{\$5,000,000} = 0.2$$

This can also be represented as

$$\frac{P}{A} = \$5,000,000 = \frac{\$1,000,000 - (1+i)^{-10}}{i}$$

From interest rate tables, for a time of 10 yr and an $A/P$ cost factor of 0.2, the interest rate is approximately 15%. [**Interest Rate Tables**]

**The answer is (C).**

**85.** Decay is an irreversible reaction. The equation for the reaction rate of an irreversible first-order reaction is

**First-Order Irreversible Reaction**
$$-dC_A/dt = k$$

Half-life is the time taken to reach half of the initial concentration. Integrating from an initial concentration of $C_0$ at a time of 0 and a concentration of $C_0/2$ at time $t_{1/2}$, the equation becomes

$$\int_{C_0}^{C} \frac{dC_A}{C_A} = \int_{t=0}^{t=t_{1/2}} -kdt$$

$$k = \frac{\ln 2}{t_{1/2}}$$

$t_{1/2} = 2$ yr, so the reaction rate constant is

$$k = \frac{\ln 2 \text{ yr}}{2 \text{ yr}} = 0.346/\text{yr} \quad (0.35/\text{yr})$$

**The answer is (A).**

**86.** Consult a table of standard oxidation potentials. Under the conditions in which standard oxidation potentials are measured, metals that are further down in the list will corrode if they are paired with metals higher up in the list. Thus, aluminum will corrode if paired with iron, zinc will corrode if paired with cadmium, palladium will corrode if paired with gold, and lead will corrode if paired with copper. [**Standard Oxidation Potentials for Corrosion Reactions**]

Note: If the conditions of the solution change, the potentials could change such that the order the metals are listed in change.

**The answer is (A).**

**87.** The structure of 3-methyl-2-butanol is

The methyl group adjacent to the alcohol bearing carbon contains carbon number one, and the alcohol bearing carbon is carbon number two. The double bond can be formed in one of two places: between carbon number one and two, or between carbon number two and three. The double bond is more likely to form between carbon number two and three because carbon number three is a 3-degree carbon and can stabilize the loss of its attached hydrogen atom by sharing the strain with the three carbons it is attached to. In other words, the most substituted alkene product is favored. The most likely product has the structure shown.

The name of this compound is 2-methyl-2-butene.

**The answer is (C).**

**88.** If addition of a nutrient increases algal growth, then it is likely that the nutrient is present in a limiting amount. When nitrogen was added, the algae in the lake grew at the same rate as when nitrogen was not added, so it is not likely that nitrogen is the limiting nutrient. However, the algae grew faster when phosphorous was added, so it is likely that phosphorous is the limiting nutrient.

**The answer is (B).**

**89.** Consult a hazardous waste compatibility chart. To determine if toxic gases can be generated by a combination of two materials, find the intersection of the two materials in this chart. If that intersection has "GT" in it, then toxic gases can be generated when the two materials are combined. All the listed pairs of materials have the potential to generate a toxic gas when combined. [**Hazardous Waste Compatibility Chart**]

**The answer is (D).**

**90.** The EPA-recommended values for amount of fish consumed by an adult is 6 g/d. [**Intake Rates—Variable Values**]

Over the course of a year, this corresponds to

$$\text{annual intake} = (\text{IR})(\text{EF})$$
$$= \left(6 \frac{\text{g}}{\text{d}}\right)\left(365 \frac{\text{d}}{\text{yr}}\right)\left(\frac{1 \text{ kg}}{1000 \text{ g}}\right)$$
$$= 2.2 \text{ kg}$$

Note: It would be incorrect to apply the EPA-recommended frequency factor of 48 events/y to the 6 g/d estimate of fish consumption.

**The answer is (A).**

**91.** Use the EPA-recommended values for estimating intake. [**Intake Rates—Variable Values**]

The available skin surface area of an adult male is 1.94 m², and the average body weight is 78 kg. The equation for the averaging time is

*Intake Rates—Variable Values*

$$\text{AT} = (\text{ED})(365 \text{ days/year})$$

The median duration of a shower is 7 min. The exposure duration appears in both the numerator and denominator of the equation for absorbed dose (AD), so it cancels out; for this example, an exposure duration of one year is chosen.

Convert units of concentration to mg/L.

$$CW = 23 \text{ ppb by mass}$$
$$= \left(\frac{23 \text{ g chloroform}}{10^9 \text{ g water}}\right)\left(1 \frac{\text{g water}}{\text{mL water}}\right)$$
$$\times \left(1000 \frac{\text{mL}}{\text{L}}\right)\left(1000 \frac{\text{mg}}{\text{g}}\right)$$
$$= 0.023 \text{ mg/L}$$

Convert the units of the available skin surface area to $cm^2$.

$$SA = 1.94 \text{ m}^2 \left(\frac{(100 \text{ cm})^2}{\text{m}^2}\right) = 19{,}400 \text{ cm}^2$$

Convert the exposure time to hours per event.

$$ET = \left(7 \frac{\text{min}}{\text{event}}\right)\left(\frac{1 \text{ h}}{60 \text{ min}}\right) = 7/60 \text{ h/event}$$

The daily absorbed dose is

Exposure

$$AD = \frac{(CW)(SA)(PC)(ET)(EF)(ED)(CF)}{(BW)(AT)}$$

$$= \frac{\left(0.023 \frac{\text{mg}}{\text{L}}\right)(19\,400 \text{ cm}^2)\left(0.06 \frac{\text{cm}}{\text{h}}\right)}{(78 \text{ kg})(1 \text{ yr})\left(365 \frac{\text{d}}{\text{yr}}\right)}$$
$$\times \left(\frac{7}{60} \frac{\text{h}}{\text{event}}\right)\left(365 \frac{\text{events}}{\text{yr}}\right)$$
$$\times (1 \text{ yr})\left(\frac{1 \text{ L}}{1000 \text{ cm}^3}\right) \left(10^6 \frac{\text{ng}}{\text{mg}}\right)$$

$$= 40 \text{ ng/kg·d}$$

**The answer is (C).**

**92.** Use the EPA-recommended values for estimating intake. The rate of swallowing water while swimming is 50 ml/h, and the average body weight of an adult male is 78 kg. [Intake Rates—Variable Values]

Exposure duration appears in both the numerator and denominator of the equation for absorbed dose (AD), so it cancels out; for ease of calculation, an exposure duration of one year is used. The equation for the averaging time is

Intake Rates—Variable Values
$$AT = (ED)(365 \text{ days/year})$$

The exposure time is 0.5 h per event.

The formula for chronic daily ingestion of water while swimming is

Exposure

$$CDI = \frac{(CW)(CR)(ET)(EF)(ED)}{(BW)(AT)}$$

Convert the units of the concentration to mg/L.

$$CW = 3 \text{ ppm by mass}$$
$$= \left(\frac{3 \text{ g chlorine}}{10^6 \text{ g water}}\right)\left(1 \frac{\text{g water}}{\text{mL water}}\right)\left(1000 \frac{\text{mL}}{\text{L}}\right)\left(1000 \frac{\text{mg}}{\text{g}}\right)$$
$$= 3 \text{ mg/L}$$

Convert the units of the contact rate to L/h.

$$CR = \left(50 \frac{\text{mL}}{\text{h}}\right)\left(\frac{1 \text{ L}}{1000 \text{ mL}}\right) = 0.050 \text{ L/h}$$

Convert the units of the exposure frequency to events/yr.

$$EF = \left(2 \frac{\text{events}}{\text{wk}}\right)\left(52 \frac{\text{wk}}{\text{yr}}\right) = 104 \text{ events/yr}$$

Calculate the chronic daily intake.

$$CDI = \frac{(CW)(CR)(ET)(EF)(ED)}{(BW)(AT)}$$

$$= \frac{\left(3 \frac{\text{mg}}{\text{L}}\right)\left(0.050 \frac{\text{L}}{\text{h}}\right)\left(0.5 \frac{\text{h}}{\text{event}}\right)}{(78 \text{ kg})(1 \text{ yr})\left(365 \frac{\text{d}}{\text{yr}}\right)}$$
$$\times \left(104 \frac{\text{events}}{\text{yr}}\right)(1 \text{ yr}) \left(10^3 \frac{\mu\text{g}}{\text{mg}}\right)$$

$$= 0.27 \text{ } \mu\text{g/kg·d}$$

**The answer is (B).**

**93.** The runoff coefficient is given, so use the rational formula.

**Rational Formula**

$$Q = CIA$$
$$= (0.45)\left(5 \;\frac{\text{in}}{\text{hr}}\right)(20 \text{ ac})$$
$$= 45 \text{ ft}^3/\text{sec}$$

Note: It is not necessary to convert units. Unit conversion gives the same result to within <1% because (3600 sec)(12 in/ft) = 43,200 in$^3$/sec, which would mean the calculation becomes

$$CIA = (0.45)\left(5 \;\frac{\text{in}}{\text{hr}}\right)(20 \text{ ac})\left(\frac{1 \text{ hr}}{3600 \text{ sec}}\right)$$
$$\times \left(\frac{1 \text{ ft}}{12 \text{ in}}\right)\left(43{,}560 \;\frac{\text{ft}^2}{\text{ac}}\right)$$
$$= 45 \text{ ft}^3/\text{sec}$$

**The answer is (D).**

**94.** Substitute the equation for area into the equation for net loss and solve.

$$L = \left(6.4 \times 10^{-5} \;\frac{\text{km}}{\text{d}}\right)A$$
$$= \left(6.4 \times 10^{-5} \;\frac{\text{km}}{\text{d}}\right)\left(\left(\frac{1.0}{\text{km}}\right)V + 7.0 \text{ km}^2\right)$$
$$= \left(6.4 \times 10^{-5} \;\frac{\text{km}}{\text{d}}\right)\left(\left(\frac{1.0}{\text{km}}\right)(14 \text{ km}^3) + 7.0 \text{ km}^2\right)$$
$$= 1.3 \times 10^{-3} \text{ km}^3/\text{d}$$

This is the net loss per day during the summer. The net loss over the course of the summer is

$$L_{\text{summer}} = \left(\frac{365 \text{ d}}{4}\right)L$$
$$= \left(\frac{365 \text{ d}}{4}\right)\left(1.3 \times 10^{-3} \;\frac{\text{km}^3}{\text{d}}\right)$$
$$= 0.12 \text{ km}^3$$

**The answer is (C).**

**95.** Substitute values into the prediction equation.

$$A = RKL_S$$
$$= (350)(0.7)(0.1)$$
$$= 24.5 \text{ short tons/ac}$$

This is the per acre value. To get the soil loss for the lot, apply the ratio of the lot surface area to the surface area of an acre in square ft.

$$\left(24.5 \;\frac{\text{short tons}}{\text{ac}}\right)(60 \text{ ft})(150 \text{ ft})\left(\frac{1 \text{ ac}}{43{,}560 \text{ ft}^2}\right) = 5.1 \text{ short tons}$$

**The answer is (B).**

**96.** The cross-sectional area of flow and the wetted perimeter must be found in order to calculate the hydraulic radius. The cross-sectional area of water flow is the area of a rectangle that has the base as one of its sides and the water level as another, plus twice the triangular area between the area of the rectangle and the side of the channel.

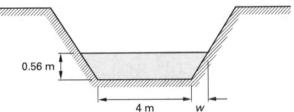

The height of the triangular areas of cross-sectional flow next to the channel sides is equal to the depth of the water, 0.56 m. The width can be found from the slope as shown.

$$\text{slope} = \frac{h}{w}$$

Solve for the width.

$$w = \frac{h}{\text{slope}}$$

Insert the equation for the width into the equation for the cross-sectional area.

$$A_{\text{cross section}} = A_{\text{rectangle}} + 2A_{\text{triangle}}$$
$$= hB + 2\left(\frac{1}{2}hw\right)$$
$$= hB + h\left(\frac{h}{\text{slope}}\right)$$
$$= (0.56 \text{ m})(4 \text{ m}) + \frac{(0.56 \text{ m})^2}{25\%}$$
$$= 3.5 \text{ m}^2$$

The portion of the wetted perimeter that is due to the sides of the channel can be found from the formula for the hypotenuse of a triangle.

$$\text{hyp} = \sqrt{w^2 + h^2} = \sqrt{\left(\frac{h}{\text{slope}}\right)^2 + h^2} = h\sqrt{\frac{1}{\text{slope}^2} + 1}$$

The wetted perimeter is

$$P = 2(\text{hyp}) + B = \left(2h\sqrt{\frac{1}{(\text{slope})^2} + 1}\right) + B$$

$$= \left((2)(0.56 \text{ m})\sqrt{\frac{1}{(25\%)^2} + 1}\right) + 4 \text{ m}$$

$$= 8.6 \text{ m}$$

The hydraulic radius is

**Flow in Noncircular Conduits**

$$R_H = \frac{\text{cross-sectional area}}{\text{wetted perimeter}}$$

$$= \frac{A_{\text{cross section}}}{P}$$

$$= \frac{3.5 \text{ m}^2}{8.6 \text{ m}}$$

$$= 0.41 \text{ m}$$

Solve the equation for predicting the length of protection for $L_P$.

$$\frac{L_P}{R_H} = 0.604 \left(\frac{R_H^{1/6}}{n}\right)$$

$$L_P = 0.604 R_H \left(\frac{R_H^{1/6}}{n}\right)$$

$$= (0.604)(0.41 \text{ m})\left(\frac{(0.41 \text{ m})^{1/6}}{0.03}\right)$$

$$= 7.1 \text{ m}$$

**The answer is (B).**

**97.** Use the equation for power to the fluid in a reel and paddle design.

**Rapid Mix and Flocculator Design**

$$P = \frac{C_D A_P \rho_f v_r^3}{2}$$

The coefficient of drag is 1.8 because the blades are flat and have a 20:1 length-width ratio. The density of water is 997 kg/m³. **[Selected Liquids and Solids]**

The equation for the effective paddle velocity is

$$v_r = v_p(\text{slip coefficient})$$

The power to the fluid is

$$P = \frac{C_D A_P \rho_f v_r^3}{2} = \frac{C_D A_P \rho_f \left(v_p(\text{slip coefficient})\right)^3}{2}$$

$$= \left(\frac{\begin{pmatrix}(1.8)(30 \text{ m}^2)\left(997 \ \dfrac{\text{kg}}{\text{m}^3}\right)\\ \times \left(\left(0.5 \ \dfrac{\text{m}}{\text{s}}\right)(0.6)\right)^3\end{pmatrix}}{2}\right)\left(\frac{1 \text{ kW}\cdot\text{s}^3}{1000 \text{ kg}\cdot\text{m}^2}\right)$$

$$= 0.73 \text{ kW}$$

**The answer is (A).**

**98.** The specific capacity of the well is

**Dupuit's Formula**

$$\text{specific capacity} = \frac{Q}{D_w}$$

$$= \frac{9.4 \ \dfrac{\text{gal}}{\text{min}}}{56 \text{ ft}}$$

$$= 0.17 \text{ gpm/ft}$$

Note: The formula for the specific capacity of a well applies to both confined and unconfined aquifers.

**The answer is (B).**

**99.** Hydraulic conductivity, Darcy velocity, and seepage velocity all have the same units of length per time. Discharge rate has units of volume per time, and is clearly not the correct answer. **[Darcy's Law]**

Hydraulic conductivity, Darcy velocity, and seepage velocity are all related to each other. The seepage velocity is the speed at which water moves through the aquifer, and is equal to the Darcy velocity, $q$, divided by the effective porosity of the aquifer, $n$.

**Darcy's Law**

$$v = q/n$$

It is faster than the Darcy velocity (defined as the flow rate through the aquifer divided by the cross-sectional area of the aquifer) because water can only move through the pores in the aquifer, not through the entire aquifer. The Darcy velocity is the product of hydraulic conductivity, $K$, and gradient, $-dh/dx$.

**The answer is (A).**

**100.** Use Theim's equation for drawdown of a confined aquifer.

**Thiem Equation**
$$Q = \frac{2\pi T(h_2 - h_1)}{\ln\left(\frac{r_2}{r_1}\right)}$$

This equation shows that if the ratio of $r_1$ to $r_2$ and $Q$ are the same, the difference in drawdown at $r_1$ and $r_2$ is inversely proportional to the transmissivity. In other words, if the pumping rate and well construction are the same,

$$\frac{Q_{well\,2}}{Q_{well\,1}} = \frac{\dfrac{2\pi T_{well\,2}(h_2-h_1)_{well\,2}}{\ln\left(\frac{r_2}{r_1}\right)_{well\,2}}}{\dfrac{2\pi T_{well\,1}(h_2-h_1)_{well\,1}}{\ln\left(\frac{r_2}{r_1}\right)_{well\,1}}} = 1 = \frac{T_{well\,2}(h_2-h_1)_{well\,2}}{T_{well\,1}(h_2-h_1)_{well\,1}}$$

Solving for $(h_2 - h_1)_{well\,2}$ yields

$$(h_2 - h_1)_{well\,2} = \frac{T_{well\,1}(h_2 - h_1)_{well\,1}}{T_{well\,2}}$$

$$= \frac{\left(1100\ \dfrac{m^2}{d}\right)(0.23\ m)}{\left(870\ \dfrac{m^2}{d}\right)}$$

$$= 0.29\ m$$

**The answer is (C).**

**101.** In a confined aquifer, the cross-sectional area of radial flow into the well is the area of a cylinder with the radius of the well and the height of the aquifer. The formula for the area of the sides of a cylinder of height $b$ and radius $r$ is

$$A = 2\pi r b$$

This is the area across which radial flow of water toward the well occurs in the aquifer. Darcy's law is

**Darcy's Law**
$$Q = -KA(dh/dx)$$

In the case of a well, radial flow is towards the well, so $x = r$ and $dh/dx$ is $-dh/dr$. Darcy's law becomes

$$Q = KA\frac{dh}{dr}$$

For a confined aquifer, substitute the formula for the area of a cylinder of height $b$, separate the variables $h$ and $r$, and integrate across the interval given in the problem statement.

$$Q = K2\pi rb\frac{dh}{dr}$$

$$\int_{r_w}^{r} \frac{dr}{r} = \frac{2\pi Kb}{Q}\int_{h_w}^{h} dh$$

$$\ln\frac{r}{r_w} = \frac{2\pi Kb}{Q}(h - h_w)$$

This is the Thiem equation, with $h_2 = h$ and $h_1 = h_w$.
**[Thiem Equation]**

Solve for $h$ and substitute values.

$$h = h_w + \frac{Q}{2\pi Kb}\ln\frac{r}{r_w}$$

$$= 7.5\ m + \frac{\left(32\ \dfrac{L}{s}\right)\left(\dfrac{1\ m^3}{1000\ L}\right)}{2\pi\left(0.028\ \dfrac{m}{s}\right)(3.2\ m)}\ln\left(\frac{15\ m}{0.15\ m}\right)$$

$$= 7.8\ m$$

**The answer is (B).**

**102.** First calculate the value for $u$, the well function argument.

$$u = \frac{r^2 S}{4Tt} = \frac{(1000\ ft)^2(5\times 10^{-4})}{(4)\left(370\ \dfrac{ft^2}{day}\right)(338\ day)} = 0.001$$

Next, calculate the terms in the well function until they become negligible.

| term | value |
|---|---|
| $-0.5772$ | $-0.5772$ |
| $-\ln u$ | $6.9$ |
| $u$ | $0.001$ |

Only the first two terms in the well function are significant. The well function becomes

$$W(u) = -0.5772 - \ln u$$

The drawdown is

$$s = \frac{QW(u)}{4\pi T}$$
$$= \frac{Q(-0.5772 - \ln u)}{4\pi T}$$
$$= \frac{\left(500\ \frac{\text{gal}}{\text{min}}\right)\left(0.134\ \frac{\text{ft}^3}{\text{gal}}\right)\left(60\ \frac{\text{min}}{\text{hr}}\right)\left(24\ \frac{\text{hr}}{\text{day}}\right)}{4\pi\left(370\ \frac{\text{ft}^2}{\text{day}}\right)}$$
$$\times(-0.5772 - \ln(0.001))$$
$$= 130\ \text{ft}$$

**The answer is (C).**

**103.** The formula for Darcy velocity is

$$v = \frac{k\rho_A g}{\mu_A}\left(\frac{dh}{dx}\right)$$

The ratio of the Darcy velocities of the two contaminants is

$$\frac{v_A}{v_B} = \frac{\frac{k\rho_A g}{\mu_A}\left(\frac{dh}{dx}\right)_A}{\frac{k\rho_B g}{\mu_B}\left(\frac{dh}{dx}\right)_B} = \frac{\frac{\rho_A}{\mu_A}}{\frac{\rho_B}{\mu_B}}$$
$$= \frac{\rho_A \mu_B}{\rho_B \mu_A}$$
$$= \frac{\left(1\ \frac{\text{g}}{\text{cm}^3}\right)\left(0.004\ \frac{\text{Pa}}{\text{s}}\right)}{\left(0.89\ \frac{\text{g}}{\text{cm}^3}\right)\left(0.001\ \frac{\text{Pa}}{\text{s}}\right)}$$
$$= 4.5$$

Solve for $v_A$.

$$v_A = 4.5 v_B$$

**The answer is (D).**

**104.** The equation for the Freundlich isotherm is

**Activated Carbon Adsorption: Freundlich Isotherm**

$$\frac{x}{m} = X = KC_e^{1/n}$$

The variables $x$ and $m$ can be flow rates. The desired concentration of benzene after treatment with carbon adsorption is $C_e$. Solve for $x$.

$$x = mKC_e^{1/n}$$

A mass balance for benzene around the carbon adsorption unit yields

benzene in = benzene out + benzene adsorbed
$$C_{\text{in}} Q_{\text{air}} = C_{\text{out}} Q_{\text{air}} + x$$

Substitute the equation for $x$ and rearrange to solve for $m$.

$$C_{\text{in}} Q_{\text{air}} = C_{\text{out}} Q_{\text{air}} + x = C_{\text{out}} Q_{\text{air}} + mKC_e^{1/n}$$
$$m = \frac{Q_{\text{air}}(C_{\text{in}} - C_{\text{out}})}{KC_e^{1/n}}$$

$$= \frac{\left(0.3\ \frac{\text{m}^3}{\text{min}}\right)\left(60\ \frac{\text{min}}{\text{h}}\right)\left(24\ \frac{\text{h}}{\text{d}}\right)}{(0.012)\left(0.02\ \frac{\text{g}}{\text{m}^3}\right)^{1/1.9}\left(1000\ \frac{\text{g}}{\text{kg}}\right)}$$
$$\times\left(\left(10\ \frac{\text{mg}}{\text{L}}\right)\left(\frac{1\ \text{g}}{1000\ \text{mg}}\right)\left(\frac{1000\ \text{L}}{\text{m}^3}\right) - 0.02\ \frac{\text{g}}{\text{m}^3}\right)$$

$$= 2827\ \text{kg/d}\quad(2800\ \text{kg/d})$$

**The answer is (C).**

**105.** This is a mass balance problem. The equation for dry solids in the stream is

$$m_{\text{dry}} = (1 - x_w) m_{\text{wet}}$$

$m_{\text{dry}}$ is the dry solid mass, $m_{\text{wet}}$ is the wet solid mass, and $x_w$ is the fraction of the wet solid mass that is water. The fraction of carbon in the dry solids is

$$f_C = \frac{m_C}{m_{\text{dry}}}$$

Solving for the mass of carbon gives

$$m_C = f_C m_{\text{dry}}$$

The fraction of carbon mass that is converted to carbon dioxide is

$$f_t = \frac{m_{C\ \text{in}\ CO_2}}{m_C}$$

Solving for $m_{C\ \text{in}\ CO2}$ gives

$$m_{C\ \text{in}\ CO_2} = f_t m_C$$

Finally, the mass of $CO_2$ corresponding to the mass of carbon converted to $CO_2$ is found based on the ratio of the molecular weight of $CO_2$ to the molecular weight of carbon.

$$m_{CO_2} = m_{C\ in\ CO_2} \frac{MW_{CO_2}}{MW_C}$$

Substitute variables and solve.

$$m_{CO_2} = f_t m_C \frac{MW_{CO_2}}{MW_C} = f_t f_C m_{dry} \frac{MW_{CO_2}}{MW_C}$$

$$= f_t f_C (1-x_w) m_{wet} \frac{MW_{CO_2}}{MW_C}$$

$$= (0.65)(0.18)(1-0.4)$$
$$\times \left(\frac{6000\ Mg}{y}\right)\left(\frac{44\ g\ CO_2}{mol}\right)\left(\frac{mol}{12\ g\ C}\right)$$
$$= 1500\ Mg/y\ CO_2$$

**The answer is (C).**

**106.** This is a mass balance problem. Express the carbon-nitrogen mass ratios mathematically. Use $x$ to represent the mass fraction of the mixed stream that is spoiled produce.

$$\frac{m_{C,produce}}{m_{N,produce}} = 18$$

$$\frac{m_{C,cardboard}}{m_{N,cardboard}} = 400$$

$$\frac{m_{C,overall}}{m_{N,overall}} = \frac{m_{C,cardboard}}{m_{N,cardboard}}(1-x) + \frac{m_{C,produce}}{m_{N,produce}}x = 30$$

Solve for $x$.

$$\frac{m_{C,cardboard}}{m_{C,cardboard}} - \frac{m_{C,cardboard}}{m_{N,cardboard}}x + \frac{m_{C,produce}}{m_{N,produce}}x = \frac{m_{C,overall}}{m_{N,overall}} - \frac{m_{C,cardboard}}{m_{N,cardboard}}x + \frac{m_{C,produce}}{m_{N,produce}}x$$

$$= \frac{\frac{m_{C,overall}}{m_{N,overall}} - \frac{m_{C,cardboard}}{m_{N,cardboard}}}{1}$$

$$x = \frac{\frac{m_{C,overall}}{m_{N,overall}} - \frac{m_{C,cardboard}}{m_{N,cardboard}}}{-\frac{m_{C,cardboard}}{m_{N,cardboard}} + \frac{m_{C,produce}}{m_{N,produce}}}$$

$$= \frac{30 - 400}{-400 + 18} \times 100\%$$
$$= 97\%$$

**The answer is (D).**

**107.** A compost pile with insufficient nitrogen is likely to heat up less than a compost pile with sufficient nitrogen. Odor problems are associated with excess nitrogen, not insufficient nitrogen. If sufficient nitrogen is present, more water is generated during composting than when nitrogen is insufficient. The microorganisms that cause composting will compost at a slower speed if nitrogen is insufficient.

**The answer is (A).**

**108.** This is a mass balance problem. The variable $m$ is the mass of aluminum over the course of the year.

$$inputs = outputs + change\ in\ storage$$

$$m_{purchase} = m_{product} + m_{scrap} + m_{landfill} + (m_{inv,final} - m_{inv,initial})$$

Solve for $m_{landfill}$.

$$m_{landfill} = m_{purchase} - m_{product} - m_{scrap} - (m_{inv,final} - m_{inv,initial})$$

$$= 1000\ kg - (1200\ parts)\left(0.5\ \frac{kg}{part}\right)$$
$$- 100\ kg - (1100\ kg - 1000\ kg)$$
$$= 200\ kg$$

**The answer is (B).**

**109.** From a hazardous waste compatibility chart, no consequences occur when inorganic fluorides are mixed with glycols, caustics, or cyanides. When inorganic fluorides are mixed with organic acids, generation of toxic gas may occur. [**Hazardous Waste Compatibility Chart**]

**The answer is (D).**

**110.** This is a mass balance problem. First, determine the concentration of carbon in the methane in units of mass of carbon per cubic meter of methane. There is one mole of carbon atoms per mole of methane, so the mass of carbon in a mole of methane is

$$\frac{m_C}{1\ mol\ methane} = \left(\frac{1\ mol\ C}{1\ mol\ methane}\right)\left(\frac{12\ g\ C}{1\ mol\ C}\right)$$
$$= 12\ g\ C/mol\ methane$$

For an ideal gas at standard temperature and pressure, there are 22.4 L of gas in a mol. [**Definitions: Chemistry**]

When natural gas usage measurements are given in terms of volume, the measurement assumes standard temperature and pressure. The equation for the mass of carbon per cubic meter becomes

$$\frac{m_C}{1\ m^3\ methane} = \left(\frac{12\ g\ C}{1\ mol\ methane}\right)\left(\frac{1\ mol}{22.4\ L}\right)\left(1000\ \frac{L}{m^3}\right)$$

The molecular weight of carbon dioxide is

$$\mathrm{MW_{CO2}} = (1)\left(\frac{12 \text{ g C}}{1 \text{ mol}}\right) + (2)\left(\frac{16 \text{ g O}}{1 \text{ mol}}\right) = 44 \text{ g/mol}$$

This means for every 12 g of carbon entering combustion, 44 g of carbon dioxide are created. The mass balance is

$$\text{carbon entering combustion} = \text{carbon exiting combustion}$$
$$= \frac{\mathrm{CO_2} \text{ exiting combustion}}{\dfrac{44 \text{ g CO}_2}{12 \text{ g C}}}$$

Solve for the carbon dioxide exiting combustion.

$$\begin{aligned}
\mathrm{CO_2} \text{ exiting combustion} &= \left(\frac{44 \text{ g CO}_2}{12 \text{ g C}}\right)(\text{carbon entering combustion}) \\
&= \left(\frac{44 \text{ g CO}_2}{12 \text{ g C}}\right)\left(\frac{12 \text{ g C}}{1 \text{ mol methane}}\right)\left(\frac{1 \text{ mol}}{22.4 \text{ L}}\right) \\
&\quad \times \left(1000 \, \frac{\text{L}}{\text{m}^3}\right)(120 \times 10^6 \text{ m}^3 \text{ natural gas}) \\
&\quad \times \left(\frac{1 \text{ Mg}}{10^6 \text{ g}}\right) \\
&= 235\,714 \text{ Mg} \quad (240\,000 \text{ Mg})
\end{aligned}$$

**The answer is (D).**